Notes from the Netshed

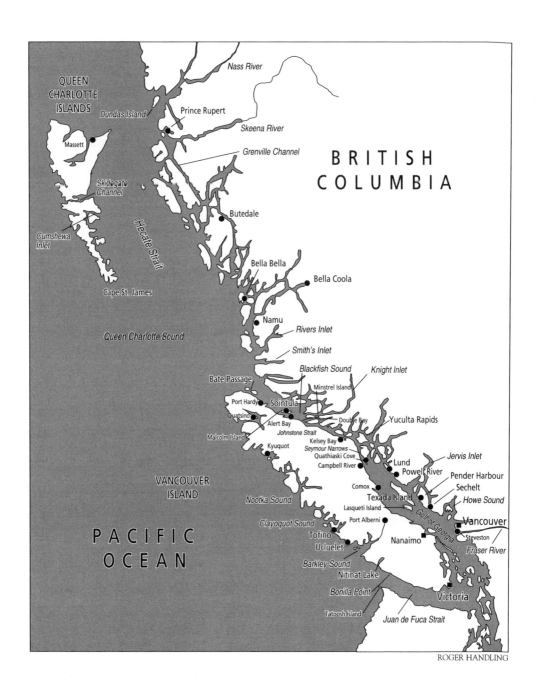

Notes from the Netshed

Mrs. Amor de Cosmos

Illustrations by Greta Guzek

HARBOUR PUBLISHING

Copyright © 1997 by Mrs. Amor de Cosmos

All rights reserved. No part of this book may be reproduced in any form by any means without the written permission of the publisher, except by a reviewer, who may quote passages in a review.

Harbour Publishing
P.O. Box 219
Madeira Park, BC V0N 2H0

Jacket design by Martin Nichols, Lionheart Graphics
Jacket painting by John Horton
Illustrations by Greta Guzek
Map by Roger Handling
Edited for the press by Peter A. Robson
Published with the assistance of the Canada Council and the Province of British Columbia through the British Columbia Arts Council.

Printed and bound in Canada.

Canadian Cataloguing in Publication Data

De Cosmos, Amor, Mrs.
    Notes from the netshed

ISBN 1-55017-172-0

1. Fisheries—British Columbia. 2. Fisheries—British Columbia—Humour. I. Title.
SH224.B7D4 1997   338.3'727'09711   C97-910836-5

*To all the women and men
in the commercial fisheries of British Columbia.
It may not always be a living, but it is a life.*

CONTENTS

Letters from Home
 David Rahn   11
Fact, Fiction & Fishing
 Vickie Jensen   13

SPRING

A Time of Renewal   17

Horizontal Rain and Ed's Deckhand   22
To Catch a Herring   26
Sex and Violence in the Gulf   30
Squid Jigging Ground   32
The Fisherman's Bookshelf   36
Brief Lives   40
Fishing for Wages   44
Yoshi   47
How to Eat a Dogfish   51
Spring Fever and the Scientific Fisherman   55
Spring Cleaning   59
The Case of the Left-handed Halibut   62
Eulachon Madness   66
The Great Eulachon Hunt   69
A Piece of Rope   72
How to Buy a Boat   76
Down and Out on Minstrel Island.   79
No Cure, No Pay   83
The Old Captain Takes a Walk in the Sunshine   86

SUMMER

The Mystery Tour	93
The Sunshiners	99
Working for the Yankee Dollar—We Build a Net	103
Two Matches and an Attitude	107
Bull and the Boys Come to Town	111
Setting Your Net	115
Adams Fish	119
Homemade Philosophers	122
Jonah in the Belly of the Whale	126
Tom the Skiffman Gets Dressed	130
Fishing in Roll-Up City	134
Fish and Berries	138
Tide and Currents	141
Turn Your Radio On	145
Fish Wars	148

FALL

Living in Equinoctial Times	155
The Secret Life of Dog Salmon	161
The Man Who Remembered Everything	165
Fisherman's Wife	170
High-Test Beer and High-Tech Fishing	174
Staying Alive	177
High Jinks on the High Seas	181
Preventive Maintenance	184
The Fisherman's Shed	188
Kiss Me Hardy, For I Am Dying	192

WINTER
On the Beach   199

The Muggins Rule   206
Dry Rot   210
The Night Pete Shot Santa Claus   213
How to Travel with a Salmon at Christmas   217
The Year of the Two Santa Clauses   221
A Fisherman's Frolic   225
The Year of the Pig   229
Roe, Roe, Roe Your Boat   233
A Good Rule of Thumb   237
Oysters and St. Valentine   241
The Wealthy Fisherman   245
A Year of Resolution   249

Afterword: Mug-Up   253

Acknowledgements   255

Letters from Home

Mrs. Amor de Cosmos first appeared in print in Ottawa in the June 1980 issue of Joe Gough's *Canadian Fishing Report*. In fifty-five "Dear Joe" letters that followed, she faithfully reported on the ways and wiles of the west coast fishery, sometimes with her tongue firmly in her cheek, always with her heart squarely on her sleeve. Throughout those years she lived and wrote about her life as a fisherman's wife, all the while remaining a mystery herself, well known in the details but enigmatic in her particulars. Then as now, she eschewed straight facts in favour of oblique references, innuendo and persuasion. It's an outlook that is immediately infectious.

Several years ago, she wrote to me: "Catching fish is easy, if the boat is well prepared, the net's in the right spot and you have fishermen's luck." She was not, of course, referring to that horseshoe, clover leaf, make-your-own kind of luck, but instead to the more mysterious union of hunter and prey that begins in the solitary dance with a partner just vaguely imagined and ends in a seductive willingness to catch and be caught.

Notes from the Netshed

When her first letter arrived on my desk in March 1991, I couldn't believe my luck. At the time I was Senior Editor of *The Westcoast Fisherman*, the country's largest and most widely read commercial fishing magazine, so my boat was well prepared and the net was cast wide. Still, the catching was a two-way agreement.

We did our dance and struck a deal—straightforward and non-negotiable: she would deliver observations and intimate details about the fishermen in her life and I in turn would make sure her words got into print, help her through the rough spots and protect her identity. I confess that she has been better at holding up her end.

In the years since our first exchange, Mrs. Amor de Cosmos has come to occupy a place of honour in *The Westcoast Fisherman*. Her monthly column "A Letter From Home" has documented not only the lives of fishermen, and fisherwomen in particular, but also the people and events that concern them most, including everything from fish prices to bad boat coffee. Along the way she has transformed herself from an observer into a fisherman. She's gone from working for others to working for herself and back again, changed partners and gear types and thrown the whole business off and taken it back too many times to count. She's also introduced an eccentric cast of dockside characters who are now so familiar they're part of the west coast family. And she's delivered some of the crispest vignettes and wittiest essays ever to grace the pages of a commercial fishing magazine. Many west coast fishermen already hold her close to their hearts, now it's your turn. Welcome to the world of Mrs. Amor de Cosmos.

DAVID RAHN
GROUP PUBLISHER, WEST COAST PUBLISHING

Fact, Fiction & Fishing

I open my copy of *Notes From the Netshed* with the same eagerness I feel for each month's "A Letter From Home" in *The Westcoast Fisherman*. I'm not alone in this enthusiasm. In fact, readers have been looking forward to Mrs. Amor de Cosmos's writing at least since 1980, when she began contributing a short column to the *Canadian Fishing Report*, an Ottawa-published paper that circulated on both coasts. When that publication folded, she started writing a monthly letter to *The Westcoast Fisherman*. All told, she's survived five editors and never missed a deadline. That kind of perseverance is like continuing to fish despite lean years, buyouts, market shifts and politicians.

So who is Mrs. Amor de Cosmos? As with many a mariner, fact and fiction easily get rolled into the same life. Whoever she is, her writing reveals someone keenly aware of fishing issues, someone comfortable with all kinds of gear and insider talk, someone intimate with the emotional roller coaster of each fishing season. Whether she is heading out on Uncle Ed's troller or trying for a job fishing dog salmon or remembering the years she worked alongside her husband (now ex-husband) on their seiner, Mrs. Amor de Cosmos is no stranger to this rapidly changing industry. She knows what she's talking about.

I like that she brings up topics like vegetarianism, porch paint vs. marine paint, the rules of cribbage, whether you should eat squid or use them for bait. This is the kind of middle-of-the-night wheelhouse talk that starts with an everyday observation and eventually spins out into a philosophy of life. It's talk that gets your thoughts going, that clears the drowsiness out of your mind like a fresh, strong cup of coffee. With Mrs. Amor de Cosmos, readers can commiserate about the lack of fish or jobs in a coming season, vent frustrations about the political decisions that rule our lives, chuckle over a few dark and salty jokes, celebrate our perseverance in spite of everything. To have her best letters collected in one book is to be invited into the world of commercial fishing and to enjoy the sense of shared community you can find there.

VICKIE JENSEN
AUTHOR OF *Saltwater Women at Work*

SPRING

INTRODUCTION

A Time of Renewal

"In the spring it is the dawn that is the most beautiful." So wrote the poet Sei Shōnagon in far-off Japan in the eleventh century. She had the time to observe her world and make notes and I am as fortunate. I observe, I make notes, and I have the time to occasionally reflect on it all. I sometimes see myself as being like her, but then again she didn't have a wrist watch. She didn't even have a bank account, so maybe I am fooling myself.

But some things are eternal. The seasons drove the life of Sei Shōnagon and to this day they drive the life of a fisherman and over the years they sure have driven me.

The astronomers know that it is the earth going around

the sun that gives us the seasons, and they have calculated the arrival of spring with exquisite accuracy. For us on the west coast of BC it is not so exact. Spring comes in fits and starts: sometimes early, sometimes late, some days with glorious sunshine and others with depressing rain. But one way or another it comes and we fishermen start our seasonal cycle. We are chained to the seasons because the fish are, and Nature's clock permeates everything we do. Longer, warmer days and more sunshine means we finally paint our boats. The season has made us feel right about what we do because it is right. We know what spring is, it is a time of renewal.

Things are never neat or exact in the life of a fisherman. A big fishery here is the herring roe fishery in March. So in February we are getting gear ready, tuning up engines, checking radars, sounders and the cursed infinite variety of radios that we all seem to have. These are the things that must be done. Every year I make lists of them. Some are always the same, such as checking that my licence fees are paid, or getting a new tide book. So while the calendar says it is still winter, we are doing very springlike things, because the calendar doesn't matter to the herring. Their calendar is that of the moon and tide.

The full moon comes at a different time every March, but it always comes, and the herring time themselves to that. We refine time to years and months and days from the sun, and with our clocks and watches we refine it further into hours and minutes. But the fish time themselves simply to the seasons and the time of the moon. We say spring comes when it is March 21. Spring for the herring is when the March moon is right. Their spring is out of phase with our calendar time—close, but not exactly.

On land, spring changes are there for all to see. In the ocean they are more subtle but just as profound. Water temperature rises a bit, and as the plankton bloom, food is generated and the ocean comes alive. The halibut slowly move out of the Gulf of Alaska and head south. The salmon feel these stirrings too, and start back toward their creeks and lakes, arriving in the summer.

Over time my springs have changed—but then so have I. And as I have changed, so have the fisheries. In the 1980s I was married and my husband and I started out on a small gillnet-troller we named the *Fantasy Isle*. It was not much, but it was ours. In those days a boat like that was all you needed to make a living because you could troll for salmon, switch to gillnetting later and rig up for longlining halibut. You had many options, even if you didn't need them all. You could make a living—you could even pay off the bank loan and eventually own the boat. So spring was exciting, because you had to get things ready for that season and for the summer season. You never knew what was in store so you prepared for every contingency. Detail, detail, detail—one can or two of canned butter, in case we run out someplace upcoast? How many spare fuses? (every damn one different) and a million other decisions.

In those days it was exciting to get up in the morning. Dawn in spring has that special feeling. The sun shines early, the air seems fresh and there is some electricity about that energizes you. Gotta get going. Go the shipyard, then to the sounder people, then to the bank, then to the other float and then back to the boat. For us troller-gillnetters, our little world was fun and exciting. It was a life. My husband and I had dreams, tempered by the reality of the work that had to be done now and all the work we had to keep doing if our dreams were to come true.

Dawn at sea was special. In the 1980s we could troll off the west coast of Vancouver Island in May. We were fishing spring salmon; there were lots of them and fish farms hadn't gone into production yet, so prices were high. At three dollars a pound, a twenty-pound fish was worth sixty dollars. No wonder we called them "smileys." If we got a few of them on the "morning bite" we smiled all day.

But it was not the fishing that made it special. It was, for me, the dawn. It was still dark when we began to fish. We had the gear in the water, trolling at a slow speed and all was quiet except for the low throb of the engine. The sun came up over the land. Soon all that was dark and black became grey and then light, but the land stayed black. The colour of the sea changed from black to grey to blue. All this happened in about an hour, and if the fish weren't biting, you had time to watch it all. If they were biting and you had to work the gear, pull up lines, change the settings, try different depths, clean the fish and ice them down—well, then dawn came all at once.

But then one year we could not go out early. More boats and fewer fish meant later fishing times. We had less fishing time. At that time we sold our little boat and bought a small salmon seiner. So into the 1990s, our spring dawns at sea were no more, simply because salmon seiners really didn't start fishing until July.

The split from my husband meant that no longer was I a boat owner with all the prestige that is supposed to go with it. But I got a herring roe licence out of it, which is kind of the same but not really. Now my spring fishing was for herring as they got ready to spawn. But it happened, as I said, in March. So what I wanted in a spring dawn only happens once in a while, when I am fishing and there is brilliant sunshine after a rain at night and lots of time to contemplate the fishing life. It was usually not to be. The reason is that herring roe fishing is so darn hectic, tense and short. One week and it's all over. In the months of April and May it is still spring to be sure, but now I am never on the fishing ground in April and May. In the 1990s we have to wait for the salmon season to open and by then it's high summer. So my springs have changed as the fishery has changed.

There is a larger lesson in all this. Spring comes in its own way, more or less the same every year, and with it come all the other spring things. There is no such thing as a perfect spring but there are some things that are eternal. The herring come to spawn and the eggs we gather go to the Japanese, who give them as gifts to symbolize fertility and renewal. A clear sign for us and for them.

And on the docks of Steveston where I live, there will be the old activities of spring renewal—painting, fixing, testing—that I have done for too many years. I will not be alone. My Uncle Ed and his pals will get their trollers ready for seasons that are now very short, and for some of them confined to their backyard, the Gulf of Georgia. Others are celebrating the success of the herring season and taking their time to get ready for the salmon season. Three months to do what used to take two weeks.

As for me and mine, well, I prowl the docks—searching for gossip, looking for the Scientific Fisherman to get the usual misinformation, and thinking of springs gone by when I was young. The world has changed, fishing has changed, I have changed but spring has not. There is always the dawn. Always constant, always renewed. I have learned that. How could Sei Shōnagon know that over a thousand years ago?

Horizontal Rain and Ed's Deckhand

January was dead. I have no recollection of doing anything or of anything getting done. But in February things sure picked up in a hurry. Uncle Ed started coming over to the house, which always perks things up, and every time he came he brought his deckhand with him.

My husband and I used to troll and gillnet, and for a number of seasons we fished out of a little wooden seiner, the *Fantasy Isle*, but some years ago we separated. Now he is a partner in an aluminum seiner which he runs, and there is a new woman in the galley. I'll spare you the details because only I know where my shoe pinches.

I tried a little schoolteaching, then opened a small gift shop—which I sold at a huge loss and was never in my life so glad to take a loss on something just to be rid of it. I've come back to the old fishing racket because it's the only thing I know and like. So I opened up all my boxes of stuff and reread all the notes I made during those many years my husband and I trolled and gillnetted. Not much on seining because we were always too busy—the less fishing time we had, the busier we got.

Now I'm Uncle Ed's bookkeeper and partner in two herring roe gillnet leases we have. Well, he has one and I have one but it's sometimes hard to get that across to Ed.

Ed has fished everything on the coast with every type of gear. But he really is a troller, who also gillnets roe herring. Ed's boat is new, has all the best gear, and he makes money. I know that because Ed brought his deckhand over for me to do the deckhand's income tax. Depending on the deal they strike with the skipper, deckhands get either 15 or 20 percent of the gross, so I have a good idea of what Ed makes through trolling, and I can guess what he makes on the herring, because we pool our licences.

A more diverse pair than Ed and his deckhand you cannot imagine. Ed is big and noisy, and very forceful. He is one of those people who need a lot of space to move around in. For Ed to say hello means that he covers about fifteen or twenty feet on either side of you, and there is much yelling and thumping on your body.

His deckhand is entirely different. He is small and quiet, and takes up no space at all. His name is Frankie, and he is a strange guy. He comes from Sointula, but he doesn't look like a Finn. He is dark, and looks like an Indian except for his blue eyes. Ed says the only thing Finnish about him is that he doesn't talk unless he has something to say, which is never.

Ed introduced him as "my Non-talking Deckhand." I never knew his name until he told me for the income tax. He was just known as Ed's deckhand.

We're busy as a bee getting ready for the roe herring fishery, and doing the deckhand's—sorry, Frankie's—income tax. Ed and I are organizing a pool on the herring, and part of the deal is that I do all the books and everyone's tax. So in spite of Ed's claim that his deckhand doesn't talk, I have actually had a number of interesting conversations with him.

Talk in our kitchen goes on at two levels: Ed talking and shouting and banging chairs and furniture, and me replying to him. We're buying nets and gear to get rigged up early, so Ed and I are hashing it out. In the meantime, Frankie just sits there. I ask him a question from time to time. His replies come whenever he has the answer ready, which may be right away or several hours later. They are invariably short and to the point.

Here's an example of what I mean. After we got the tax stuff done, and Frankie had become part of the furniture around the house, I casually asked him why he left Sointula. He inhaled. Then he exhaled. Then he was quiet. Finally he said two words, "horizontal rain," and said no more.

I knew better than to press him. That night in bed I pondered his answer. What the hell is horizontal rain? I was determined to get to the solution, but I knew I wouldn't get it from Frankie. So when I next got Ed in a quiet spot, I laid it all out. My question, Frankie's answer. Ed snorted. "Simple," he said.

The explanation is worthy of being a Zen riddle. It seems that in Sointula, Frankie lived far from the centre of town where the school, post office, and co-op store were. So for Frankie to go to school or do anything, he had to walk to the centre of town and then walk back home. A fair distance, according to Ed. Well, in the winter the wind blows from about October to March at about a steady thirty to thirty-five miles an hour southeasterly. And it's always raining. Therefore, the rain comes at you horizontally. Frankie didn't like it, so he left.

So it's herring roe time again. I'm a partner in the pool

with Ed. I know his deckhand's name and why he left Sointula. I can carry on a conversation with Ed and Frankie both at the same time, like talking on two channels at once. I am busy, and doggone it, I still get a big kick out of the fishing business. I mean, where else could you learn about income tax and horizontal rain at the same time?

To Catch a Herring

The roe herring fishery is upon us, and Uncle Ed and I and our crew will try to make some money as we sorely need it. Ed's troller will be the hotel for the six of us and it will be, as always, a tight squeeze, as we are operating two punts.

How we catch herring got me to thinking about the various ways there are to catch a herring. I was over at Ed's one night and he opened a can of sardines—which are really young Bay of Fundy herring—and put them on toast and proceeded to eat them. Ed has a funny habit of justifying everything he does. One of his little sayings is that whatever he is doing is "one of the last remaining features of an educated man" or words to that effect. When he switched from drinking rye whisky and ginger ale to gin and tonic he used to say that drinking tonic water was "one of the few remaining links among the English-speaking races." Eating sardines on toast late at night was "one of the few characteristics of a classy person." Or at least so says Ed.

Anyway, that got me thinking about sardines and how

they catch them. In the Atlantic provinces and the USA they use weirs, which are a special kind of fish trap. It's a neat system, as the herring can be held until the fish plants are ready for them. We kind of do that here in BC with our roe on kelp operations. They catch the herring in a seine and then tow the bag of fish to a pond that has kelp streamers in it. The herring spawn on the kelp and the Japanese pay big bucks for this delicacy. As the man said when he kissed the cow, it's all a matter of taste.

In BC we have two other methods of catching herring. One is old and one is new but like everything in the fishing industry, new things are really old ones in a new suit of clothes. The herring seiners catch herring the way they have always done and those who remember the reduction, or fish meal, industry would see nothing new in how they fish.

In the old days they often pit-lamped the herring, using lights to attract them, and the fishery lasted for months. The industry now is a gold rush where the fishery may last only fifteen minutes. But the technique is the same: set the net around the herring, purse it up, and brail or pump the fish aboard. They use nylon nets instead of cotton and nylon lines instead of manila, but otherwise there are no fundamental changes. Well, maybe the drum seine technique is a change from the table seine, but there a crewman would take about ten minutes to adapt from the old to the new.

Not so with the herring gillnet industry here. It's all new (except that it's old). The advent of the roe herring fishery saw the rise of a gillnet fishery. The use of gillnets has a special reason. The females, being full of roe, are generally larger than the males, so a gillnet will get more females than males. For those who don't know, a gillnet catches the herring by the gills—big ones get caught, small ones can get through.

Once the herring are gilled, they must be removed from the net. Salmon are picked out of the gillnet by hand, one at a time. This wouldn't do in the herring fishery. There are about 3,000 herring in a tonne. If you dream of a 5-tonne set, then you have to get 15,000 of them out. Picking them out one at a time would be a nightmare. So BC fishermen invented an ingenious device, a "beater" to shake the herring out as the net comes aboard.

On a herring punt the net comes aboard amidships. The beater bar beats the net and out drop the herring into the pens. A few still have to be picked out by hand, but the beater bar solves a big problem.

The herring fishery in BC has had a number of changes over the years. In the 1920s and '30s they were salted for the Japanese market. Then came the reduction fishery for fish meal and oil, and the herring were almost wiped out.

The new roe fishery got started in the 1970s and that is the fishery we have now. But it is changing: the herring stocks are down and the licensing restrictions are getting tougher and cost more. But what can we do? All the fisheries are regulated and before you can put a line in the water you have to have several licences, certificates that your boat is seaworthy, survival gear, and a good banker behind you. I knew there would be many tough days in the fishery, but not this many.

Well, I am not going to complain. There are many crybabies in the industry, but also there are many who can accept it all and get on with life. Things are so tense in the industry that the voices of moderation are not heard. We all know that there is a need for a fundamental change, it's just that none of us know what it should be.

There are a few laughs to be had. One of our crew is somewhat religious and packs a Bible. One of the new guys said that he shouldn't believe anything in the New Testament because most of the witnesses to the major events were all fishermen. We aren't atheists but we all had a laugh. We need it. Fishing herring is no laughing matter.

Sex and Violence in the Gulf

"Sex and violence in the Gulf, eh Uncle Ed?" I said as I went into the living room where the TV set was blaring out a documentary about Saddam Hussein. Ed glared, as I had found him out. He always says that TV only has sex and violence and high-speed car crashes on it so he never watches it and here I had him red-handed. I knew he would say something to deny he was watching it or would try to change the subject by attacking me upon some pretext or other. Not only does Ed not like to lose, he never admits that he has—on anything. I braced myself for the storm.

"The only damn sex and violence is in the roe fishery in the Gulf of Georgia!" he shouted at me. I had a vague idea what he was talking about so I got ready for a long, drawn-out analysis of the roe herring fishery that may or may not be based upon fact but certainly would conclude with him being right and everyone else completely in the wrong. I once told him he was often in error but never in doubt. He took it as a compliment until he figured out that it was not

Sex and Violence in the Gulf

complimentary, whereupon he completely forgot that I ever said it. Ed either doesn't notice criticism or he uses every weapon at his disposal to cripple his antagonist. But I wanted to hear about the Gulf and the fishery.

Ed's steamed about the price we're going to get for our fish. We were going to lease out our licences but the bottom has fallen out of that market and no one wanted to give us up-front cash. So we tried to put together a deal where we would use Ed's troller as a hotel and manage two punts. Well, getting new partners is worse than getting a new husband or wife. Deals kept crashing all around us. In spite of what Ed says it wasn't sex and violence. It was about money. I fired up my calculator and my computer and I cranked out so many different spreadsheets under different assumptions that I even confused myself. Ed and I were in a jam. We had two licences and herring punts and his troller, but no shakers to crew the punts and no deals with any buyers. And the cash buyers we tried to link up with, they all got cutesee-wootsee on me and wouldn't even admit that it was Thursday. On top of it all, the sun was shining and the water temperature in the Gulf was high and this meant the fishery could start early and we'd still be tied up in Steveston.

Well, as it turns out just as things are at their blackest, blue skies can appear. While Ed was giving me hell about our collapsed deals the phone rang. Three hours later it was all sewed up. I had hired the shakers and we got a cash buyer whose price was livable. We were ready for the Gulf of Georgia herring roe fishery.

When I asked Ed if we should go for the deal he said, "Take it. Cash is king in this business." I said, "No Ed, cash is *queen*!" "Oh," he said, "more damn sex and violence, eh?" It didn't make much sense to me but from his smile I knew it meant a lot to Ed.

Squid Jigging Ground

One of the features, if not one of the joys, of being in the fishing racket is that you have lots of time to talk. That is, if you want to, you certainly have the opportunity. This March, we were in Courtenay waiting for an opening in the Gulf of Georgia to fish for herring and their prized roe. As usual, we were working two punts and using Uncle Ed's troller as a hotel.

Before we went out onto the grounds and anchored up at our old spot, we got everyone together in a parking lot that was convenient to the shaker crew. Organizing cars and pickups and gear means you need a parking lot. Funny thing, to get four shakers, Ed and me all together at the same time took three pickups and two cars.

Everything worked out okay. This being business, with a little help from cell phones we were all there at the appointed time. All of us showing up at once was such an impressive spectacle that one of the shakers volunteered to buy lunch and we ended up in a seafood restaurant—where

Squid Jigging Ground

else? We all ordered beer except for one guy who ordered a rum and Coke. We then ordered our food and I asked for the calamari, which to you and me is squid. The rum and Coke guy disdainfully remarked, "You eat those things? Where I come from we use them for bait." This started a long discussion about squid and fishing in general. For you see, that is how you learn about fishing—no books will give you the actual knowledge you need to get by.

While the sex life of squid is remarkably boring, unlike that of its cousin the octopus, they are very interesting otherwise. People have been eating them since the dawn of history at least. The Mediterranean style of deep-frying them is the most popular here, although I have also had them stuffed and baked in a tomato sauce. They are usually served as a starter and appear on the menu as "calamari." I really like them, and with a bit of pita bread and hummus dip, I consider that a meal.

But I am meandering. The rum and Coke guy was from Cape Breton Island and said that when he was a kid, fishermen in the small boat fishery would jig for squid and use them for bait in the longline fishery. They definitely didn't eat them. The squid we eat here on the coast come from California. If you read the labels on the frozen package in the supermarket, which I do, most of it comes from the Monterey area and is quite cheap. I happen to know how they catch them. They use small seines just like we fish for salmon and they use a power block as in Alaska seiners. I hear that around the San Francisco area they use the lampara seine, which is a small seine that doesn't purse up. It merely collects the squid in a bag in the middle. The rum and Coke guy told us about how to jig for squid which is a no-brainer. Throw the line over and jig. The jig is your standard model, a weight with a treble hook attached.

33

While we were on our third beer (or was it our fourth?) I told them about the time when I was married and my husband and I had our small seiner. We had ripped up our net the week before and just laced it up. So while we were anchored up, waiting for the opening, we backhauled the net and worked on the rip. We worked until dark because we wanted it done. We had left the decklights on and when I went out on deck around midnight I looked in the water. To my surprise and delight there they were: thousands of squid. They were attracted to the lights and were flashing away in the water. That's how they catch them in many parts of the world. They turn on the lights and usually have some sort of a brailing or scooping rig to get them out of the water.

Sitting in that restaurant in Courtenay we may have looked like ordinary working people: four nondescript guys, an old beat-up looking one (Uncle Ed) and this woman (me). But it was a school on the catching of squid around the world. Jigs, seines, lampara nets, lights, and brailers were just some of the techniques discussed. To me, fascinating knowledge. We kind of had to defer to the rum and Coke guy because none of us had actually fished for squid and I was probably the only one who eats them on a regular basis. When we finished, our minds turned to herring, but we had learned a lot about squid.

The Lord knows how much money we will make on the herring. The talk is of the weak yen and the strong dollar and the effect it will have on our prices. Down and dirty. We dare not talk of the salmon season as it is always chaotic and predictions are useless. But these are tough times for me as the industry is shrinking and it is becoming necessary for me to find a new fishery. Times are so bad that I have started thinking the unthinkable. I had an offer of a job in a sports fishing camp. It was as a cook and I won't go that route. But

I was tempted. I mean I could guide, or at least look after the equipment at the docks or whatever, but it goes against my grain somehow. I want to be in the old-time fisheries—which of course are no longer the same.

If I were a realist I would note that there are fewer jobs in the salmon and herring fishery, Uncle Ed is ready for retirement as he is an old troller and the younger guys are into many things other than fishing. But because I am a bit of an optimist (I need to be!) I am sticking in the industry. Maybe I'll try and get into the squid fishery which I hear is starting up. I know a spot where they can be found. I also know lots about catching them and I do eat them so that makes me an expert. Also, I know about their sex life. Pretty boring, but don't take my word for it, go look it up.

The Fisherman's Bookshelf

Herring season is upon us. As the owner of a single Gulf of Georgia gillnet herring roe licence I will be staying on Uncle Ed's troller while my crew fish my punt. The usual cold fear of not getting any fish is joined with the excitement of possibly getting lots. These emotions can stir my soul. This fishery only lasts a few days but it is my major source of income. My bookkeeping and income tax business keeps some money flowing in and the odd job on the salmon allows my body and soul to remain together. So far.

But so much fishing these days is really just waiting around. We anchor up for days waiting for the opening. This means after you get tired of everything else you turn to your bookshelf to improve your mind or at least divert it.

What is on the fisherman's bookshelf? Well, I guess it varies. One book that is sure to be read is the yearly tide book or formally *Canadian Tide and Current Tables*. But as one wag put it, the book is interesting but the stories are too short. And only a rocket scientist can figure out its explanation of what causes tides and currents. The other well-read book is the venerable *Captain Lillie's Coast Guide*. It is the cat's meow on how to get from A to B on the coast. Some boats even have the "BC Pilot," also known as *Sailing Directions—BC Coast*. These books are necessary tools of the trade, but of course radar and the GPS satellite navigation are making them obsolete on the bigger boats.

But some books are read for pleasure. The best books are the ones you can read and then re-read for new pleasure. One book, *Chapelle's Boatbuilding*, is well thumbed. It is about American boats built in the twenties and thirties. Not many of them are fishboats, but it still is a good source if you are into how wooden boats are built and why they are the way they are. There is just no one as good as him at explaining what a "stopwater" is and how it should be placed. You can read *Chapelle* and in your mind's eye compare it to a boat you know and you can get into the thinking of its boatbuilder. Believe me, it's fun.

One book that gives great pleasure and is also a capsule history is Walbran's *British Columbia Coast Names: Their Origin and History*, published oh so many years ago by a true sea dog and amateur historian. He sailed the BC coast at the turn of the century on the government vessel *Quadra* and probably visited every place he wrote about. He named a few spots also—Florence Island in Rivers Inlet is named after his sister. Every entry seems to contain a little nugget of information. I never tire reading him. In Walbran's book you also get a sense of the English upper class who had sons in the navy and who surveyed the BC coast. Britain really did rule the waves in those days.

Then there are two books I have had for years and just can't seem to get through. I guess every bookshelf has to have unread books. One is *Moby Dick* which is just too long, even though its opening sentence is a grabber: "Call me Ishmael." What a line. The other I bought at a garage sale and thought it was the key to the fishing industry. Maybe it is, because I can hardly make head or tail of it. It is the Bible of fisheries biologists: *Computation and Interpretation of Biological Statistics of Fish Populations: Bulletin 198* by Ricker. I've tried and tried but it's no use; I still have no idea how they estimate salmon stocks. The only thing I think about is that I once was in the Legion in Alert Bay and it had "198" in big letters in the tiles on the dance floor, those numbers being the Legion's branch number. The significance of this escapes me.

I can cook as well as or better than the next person, but I have to have cookbooks. I have a dog-eared copy of *Fannie Farmer* and the *Joy of Cooking* and for exotic recipes and pictures of fish I have never eaten I refer to McClane's *Encyclopaedia of Fish Cookery*. Somewhere I have a box of cookbooks and I see that the bookstores always have new ones for every taste, but these three are enough for any fisherman. Some cooks I know disdain their use and pretend that they have never read one, but I think that everyone has a *Fannie Farmer* somewhere.

Now these are my books; I am sure every fisherman's bookshelf is different than mine. Some are into travel books, some into gardening books and others into you-name-it. Reading is a personal thing. Late at night in the privacy of your room or bunk, it lets you escape the drudgery or anxiety of this fishing life. It is a life that has many parts: fish, money, work, people and dreams. We may never realize our dreams, but through our books we can dream, sometimes with the author and sometimes on our own. When an engine won't start we sometimes use starter fluid to get it going. When we are dispirited and need a vision or a dream we need a book. Books are starter fluid for the mind. I swear that someday soon I will once again tackle *Bulletin 198*.

Brief Lives

Everybody knows that March either comes in like a lamb and goes out like a lion or vice versa, but only fishermen know that March is herring month. So March is when you find me out on the grounds. Well actually, out in the Gulf of Georgia as that is the only place I can fish with my licence and my punt.

The crew and I live on Ed's troller and every year it seems to get more cramped. Part of the reason is that Ed sits at the galley table all the time playing 714 games of solitaire a day, drinking foul instant coffee, and snorting at what he hears on the radio. I know this as I am sitting there with him. I keep updating my journal and notebooks. When I first started fishing I thought of a university career and being a writer. So I have a box full of journals and notes. Sadly, most of it is useless I guess. I mean, notes on how to fish in areas that are now closed aren't of much value.

"What the hell are you writing about now?" Ed will ask me when he looks up from solitaire game number 715. I

Brief Lives

usually give a mumbled reply or say simply, "Stuff." Herring fishing can be boring as you wait for the opening and to pass the time I write down the name of the boats I see and maybe a little note about the people. The notes about the people I call "Brief Lives."

One of the curious things about the fishing industry is that you meet people only briefly. You see the boats again and again and come to know more about each boat's history than you do about its owner and his or her life. My notes about a boat may be long but those about its owner very short. And that short bit may only be about some fishing thing, but sometimes it records some of the owner's quirks and kinks. Cruel lies, fiction, slander, and character assassination are part of the fisherman's knowledge about people on other boats. For example, if you say the name of a certain boat everyone thinks "Pruneface" because the skipper has a very wrinkled and weather-beaten face. You get the idea.

Here, from my notes, are some brief lives:

"Scotty"—owns a salmon gillnetter which is neat as a pin but has nothing new. Does not drink or smoke and is very generous with his time and his advice. Always willing to help. Would not lend you a dime. His Scottish ancestry somehow made him what he is. I think he has lots of money but as Ed says, no one has ever seen it. What good is his money if there is no pleasure in it? He likes Milbanke Sound as his conversation is always full of it.

"Frank"—a most curious man. In his middle sixties but seems ageless. Related to an engineer we had on our seiner. Lucky in fishing, unlucky in his choice of wives. His secret hobby is painting. Owns a small seiner which has had the same crew for eons. Has one blue eye and one brownish. Can also play the violin but what tunes or how well I do not know. People like him but for what reason again I do not know. Over the years, Frank has not changed in appearance and neither has his boat. The question is: does a man get his personality from his boat? Or is it the other way around?

"Mike"—a troller and a vegetarian. Always has crew problems because of it. How did he arrive at that situation? A mystery. But then, there are many in life and in the fishing industry.

"Sally"—an early widow who found a man willing to take her and her children. Her husband refers to it all as a "ready-made" family. They are very happy but the children have caused them much grief. They have found joy in each other and this is good, for who amongst us can say he is really happy? In my heart, I believe they are, in spite of much sadness in their life. Their boat is a gillnetter and they are always together. I have never seen him without her.

"Brian"—a Native Indian successful in fishing and in the business of licences, although he wryly points out that many of his relatives are not. Very generous to his relatives, although I have heard him say "Why can't I have one rich relative?" Married very young and a grandfather before he was forty. A good dresser: wears a suit and tie frequently, which is strange in the world of fishing as I know it. Very good with knots and with the splicing and mending of nets which comes from his experience as a trawler—or dragger, as he would say. Uses a strange turn of phrase to sum things up: "It makes no never mind."

"Pete"—I think he is Dutch or German. He has the look of a person who has been trained in a skill and has that quiet confidence of people who get things done. Very measured in his speech and in his actions. When I think of him, I mentally say "no fuss, no muss." I am sure that is how his life is.

These "Brief Lives" are not much, but they are the total of what I know about these people. Others know them better, others less. The fishing industry sees many people come and go and seems to welcome them all. Some stay, others leave. Where they go I don't know, because this is my life and here I am. I often muse about what other people might say about me and how much they really know about my life.

There is no sense pursuing this with Ed as I know his reply would be that it doesn't matter and once you're dead it also doesn't matter. But what of my notes? To quote an old author: "These things represent the eye, the places, customs and fashions, that were of old times." These "Brief Lives" may do that. In the meantime, the herring don't care. Their brief lives are all the same, just as all fishermen's lives may be different but they are the same when it comes to fish.

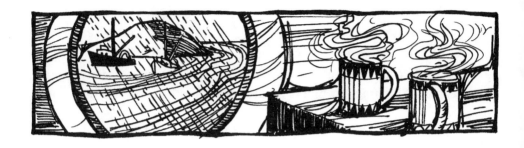

Fishing for Wages

Getting ready to fish the roe herring in the Gulf, I knew that me and Ed and the four shakers using Ed's troller as a hotel would have some adjustment problems. So I did my planning. The major plan was to anticipate Ed and get everything tiddly beforehand. I worried because I do lots of things that irritate Ed and he has lots of habits that irritate me. One big one is the fact that Ed always says or does the same things in certain situations, and that somehow gets on my nerves.

For example, Ed has standard replies for people who ask him how much money fishermen like him make. He has three answers. "Just wages" means Ed doesn't want to tell anything. "Less than wages" means Ed has had a tough time. Number three, "better than wages," is used rarely. This probably means a load of sockeye.

The major irritant I was worrying about was the fact that Ed has a little thing with his coffee that drives me up the wall. After stirring it with a spoon he raps the spoon on

Fishing for Wages

the side of the cup to get the coffee off of it. The ching! ching! noise is like touching a sore tooth.

This all came up when I asked him what kind of coffee I should get for the boat. I have one of those fancy little glass and filter rigs that all the yuppies make their coffee with. The shakers, on the other hand, would probably want your basic fishboat coffee. Ed said he didn't care what kind but to also get lots of instant. I was the cook and I wanted a happy crew. Three kinds of coffee it was. No problem.

Fishing roe herring is lots of waiting, listening to the radio and coffee drinking. We went to our old spot by Denman Island and stood by. Radio drove me nuts. I made my coffee in what Ed came to call my high-tech pot. He drank instant coffee and drove me nuts with the spoon routine—ching! ching! ching! it was all day long. For the shakers, I made fishboat coffee. Here is the recipe: *Put one spoonful of coffee in the pot for each cup, plus one for the pot. Add cold water and put it on the stove. When the water boils it is done. Let it settle and then drink it with Pacific canned milk.* We still tie it down on the stove with that little chain and lead weight thingy that has no name but has been around forever.

For my notebooks on fishing, I wrote all I could on how the herring punts have changed in design and how the nets have been modified since I fished herring years ago. But all that will have to wait for the day I have time to write it up.

The opening came quickly. Ed calls this kind of fishing a "no-brainer" as while you select with some care where you set your net, all you can do after is wait.

And wait we did. The coffee consumption increased. Then we hauled. Then we waited for the cash buyer. Then we set again. The joy of fishing is there, and I find it hard to explain, but it sure is great when you get those pens filled with fish. At this point you sometimes see a change in the personality of people: Ed gets crankier, I get happier and one of the shakers who never said anything suddenly bursts out into snatches of songs from operas.

The Gulf fishery was very good for us. Weather was also first class. We got our share, as they say, but it was hard work for the shakers. More fish means more work, but it also means more money. Cash buyers pay off in hundred-dollar bills, and every night I slept on a pile of them.

Then suddenly it was all over. Into Nanaimo, put the punts on a trailer for storage, put the nets into the net loft lockers, tie up Ed's troller and catch the ferry to Vancouver. I had to do the settling up. No problem with my computer and calculator. Paid all the bills and got the cheques ready for the shakers. They were happy.

A few days later Ed came over to get his share of the catch money. He had the troller so he ended up with more than I did, even though we pooled our licences. I didn't mind, but I was thinking of the coffee spoon routine and kept saying "ching! ching! ching!" to myself all day long. Anyways, he came in making all the noise he could, which is his style. I couldn't help myself and made a point of offering him coffee. He said he didn't want any of that high-tech stuff and asked for instant. I said I didn't have any. Nothing else to do but get to the business. I showed Ed the accounts, at which he merely glanced. I counted out his share in hundreds in little piles on the table. I couldn't help myself. "Here's some coffee money, Ed," I said.

"Just wages," he said, "just wages," and then put the cash in his pocket. At that moment the spring herring fishing was officially over.

"See you kiddo," Ed said, and he left, whistling off-key as usual.

Yoshi

Roe herring season begins around March the first, which according to my calendar is St. David's Day—the patron saint of Welshmen. I don't know if there are any Welshmen in the herring fishery but there is just about every race and kind of person in it.

One of the peculiarities of the fishery is that you spend almost all your time waiting for an opening. Fishing and getting the money are easy but the waiting is murder. You review your life up to this point, think dark thoughts, and have fantasies about making it big. That way lies madness, except that once you get fishing it all evaporates and you're back in the here and now.

This season I got to thinking about a guy called Yoshi. From his name you can guess he is Japanese. Well, Japanese Canadian, because he was born in Steveston about 1930 or so.

I know all this because of Valerie, who is related to Uncle Ed's first wife and lives on Vancouver Island.

Notes from the Netshed

Whenever she's in Vancouver she comes down and visits us. Valerie keeps us up to date on all the Island people and their goings-on. She is married to a fisherman who fishes on his father's herring seiner. Valerie has kind of taken over the role of busybody and manager of her father-in-law's boat, or so she thinks.

Last year Valerie and her group of confidants got together and decided who was to be the new cook on the boat. There is an intricate web of relationships behind getting one of these valuable jobs and Valerie thought she had it all worked out. Her candidate for the cook's job was a local guy and she figured it was a done deal.

It all blew up in her face. The captain calmly announced that the cook for the herring fishery was to be Yoshi.

Who the hell was Yoshi? No one had ever heard of him. They figured out he was Japanese Canadian because of his name, but they had to do a lot of prying and asking around before some facts began to emerge.

Yoshi had grown up and gone to school with the captain in a fishing village on the Island, where his family owned a small shipyard. But in 1941, Yoshi and his family were taken away, as were all the Japanese Canadians on the coast. Yoshi ended up in Toronto after the war, married and had kids. It seemed that he would never see the coast again. But in life, things are not always predictable or fair. The recession came, Yoshi lost his business, his wife divorced him and in his sixties he was another cold statistic among the unemployed. So he returned and asked his old schoolmate and boyhood friend for a job. And he got it.

As all these facts emerged, Valerie was beside herself. Not only had her choice been rejected, but Yoshi, the captain's choice, had never fished. One thing about most herring skippers is that they are used to making decisions, and

Yoshi

once having made them will stand for no criticism. Criticism is not a problem for them. They really believe that no one could ever even think of doing it. So the season began and the crew had to get used to Yoshi and he to them. Valerie was left on shore to stew about it all.

When the season was over, Valerie pressed her husband for details. The boat had had a good season, but it turned out that Yoshi, a failure in many things, was also a failure as a fisherman and as a cook.

Valerie's worst fears or deepest hopes were realized. But the injustice of it all burned her. The boat's success and her husband's substantial earnings were a mixed blessing, as Yoshi also got his share, the same as her husband's.

Valerie revealed all this last fall to Ed and me in Ed's kitchen while Ed sat there and made no comment and I kept the coffee flowing and tried to make sympathetic noises.

How could her father-in-law hire such a deadhead? As for Yoshi—had he no shame? He knew he couldn't cook and yet he took the job. Worse yet, no one had the gumption to confront her father-in-law or Yoshi and spell it all out. Valerie was steamed. She was used to having her way. A herring job given not only to a stranger, but to a stranger who was a loser! It was too much for her and none of us could offer her any advice.

Frankly, my heart went out to this guy who Valerie seemed to think had ruined her life. It sounded like the job her father-in-law had handed out to Yoshi was all that had kept him from hitting rock bottom.

It must have at least ruined Valerie's winter, for when I phoned her at Christmas on another matter she brought up the matter of Yoshi again. She was still mad about it. A couple of weeks ago she phoned. She and her husband were coming to Richmond and would be at Ed's. Could I bring over

49

those books that we'd talked about? Of course I could, knowing I was being set up to hear more of the Yoshi business.

When I went in to the kitchen—fishermen don't sit in the living room except for hockey games—Valerie was grinning as if she had won the lottery. It was too much for me, so I came right out and asked. Where was Yoshi, with the new season approaching?

Valerie told me the news. Yoshi was not in the crew any more. His career as a fisherman had ended.

My heart sank. The rest of us had spent all the money we made on herring months ago. What was going to become of Yoshi?

Then I heard the whole story. Yoshi had not spent any of his herring money. He had held onto it until he finally found what he wanted: a small shipyard on one of the islands that was for sale cheap. "Yoshi's going to Jones Cove," said Valerie, sounding as if Public Enemy No. 1 was being sent away for good.

"If you say so, Valerie," I said, "but you know, it sounds to me more like Yoshi's coming home."

How to Eat a Dogfish

There are two parts to fishing herring. Part One is the fishing itself and all its problems. Where and when should we set? Will the fish come? How many will we get? Part Two is the money side. Will we get a bonus? What will be the final price? And as always, what will we do with the money?

Last time we went out, we took a chance and didn't upgrade our gear. No new motors, just new ignition systems on the outboards, slightly used nets and no maintenance on Uncle Ed's boat. This violates our principles, but cost cutting is the order of the day. It's okay if everything works, but I hate that uneasy feeling of knowing that your gear is not first class.

In the end we got away with it: no breakdowns and some fish. For the record, there are herring in the Gulf of Georgia. But when you are out there, you can't help but notice other fish and sea life. Some sports-fish guys came alongside and shot the breeze with us. They were getting winter springs which to my mind are odd fish as they come at the wrong (to me) time of year. Also, they seem to be softer and whiter than summer springs.

Now, salmon sports fishermen and commercial herring fishermen usually don't see eye to eye. The reasons are obvious: they generally want the commercial guys out of the picture, and think we are killing off the herring so that there will be no more salmon. We feel they don't know what they are talking about—we weren't the ones who cemented over all the creeks to build subdivisions. Anyhow, the differences are too touchy so no one brings them up, leaving it to our respective mouthpieces to make their points.

The sports guys and us chatted away and found that we could agree on one thing: we hate dogfish. Pure and simple, we hate them. We hate them because for us they have no commercial value, they are just a nuisance. The same goes for the sportsmen—they want nice salmon on their lines, and dogfish just ain't the same.

I guess the main reason we don't like them is because technically, the spiny dogfish is a type of shark. No one knows or cares about its Latin name. Being a shark it has a skin like sandpaper, ugly green eyes, a spine in its dorsal fin that can stab you and a number of other nasty things, such as the fact that dogfish don't lay eggs. Their young are born live. Somehow, this seems like a disgusting thing for a fish to do. The list goes on. I guess after sea lions and seals, dogfish are the next most hated by us fishermen. It was not always the case.

Old-timers will tell you that during World War Two there was a big fishery for dogfish, using sunken gillnets. The livers were the objective, being very large and rich in vitamin A. Sometimes peering into old net lockers you will see the square galvanized cans that the livers were put in. They have no other use that I know of, but they are still around after fifty years. Synthetic vitamin A put the end to that fishery. Now and then people here still fish for dogfish on a commercial basis but it appears to be hard work for few dollars.

Shark is eaten almost everywhere in the world and dogfish are a big item in England in the fish and chips industry. They are sold as "rock salmon," of all things. One of the sports guys was quite knowledgeable about all this. He told us how to clean a dogfish so that you can eat it. Some of our crew listened in fascinated horror at the very idea of cleaning and eating a dogfish, but it can be done. Here's how:

First, a dogfish must be cleaned as quickly as possible once it is out of the water. Chopping off its tail and letting it bleed a bit in the water is an accepted method. Next, cut off the fins and make circular cuts around the head and tail and a cut along the back from the head to the tail. A good set of pliers is then used to pull off the skin. This is just the start. Dogfish don't have bones, just cartilage, so filleting just means running a knife along the backbone. These are the fillets you want.

Now the next part is important. Dogfish don't urinate, I am told. They excrete urea through their skin, therefore there is lots of urea in their flesh and if you don't get rid of it, it becomes ammonia and you have the wet diaper smell that will put you off eating dogfish. Most soak the flesh in vinegar or lemon juice. You can also put it in a brine solution—or even milk, I have heard. You are now ready to make fish and chips.

Uncle Ed listened to all this from the sports guys and then added that he had a favourite recipe for cooking dogfish. We all listened, out of respect for Ed's age and experience.

Ed's method was to soak the dogfish in milk, then place it on a wooden plank and squeeze a lemon over it. Add ground black pepper, some good olive oil and then, surprise!—using a heavy-duty stapler, staple the fish to the plank and put it in the oven at about 350 degrees. After about twenty minutes, take it out of the oven, throw away the dogfish and eat the plank, being careful to spit out the staples. "YOU CAN'T EAT DOGFISH!" Ed roared at the end of his story. We all laughed, even though I had heard it before.

One of the sports fishermen got off the last shot, noting that sports fishermen were considered well-known prevaricators in matters of fish, but he said us commercial guys were in a class above them. Ed beamed. In fact he beamed all day, pointing out how he duped the sports guys who were gullible, easily fooled and generally inferior to excellent personages such as himself. I didn't bother to tell him that the dogfish belly flaps are smoked and eaten in Germany as *kalbfisch* and that the fins are used in sharkfin soup and that the skin can be made into leather. Only one question remains: you know how to eat a dogfish—but will you? I know Ed won't, even though he knows how to cook them.

Spring Fever and the Scientific Fisherman

Spring fever comes in the springtime, but spring comes at different times to different people, if you know what I mean. Obviously, there is the calendar and the seasons more or less follow it, allowing for the stuff in the *Farmer's Almanac*. But spring is also a state of mind.

Politicking is a winter sport and herring fishing is a springtime thing. But one year, we fishermen were in the middle of meeting with politicians and so forth to decide our futures when all of a sudden the Gulf of Georgia roe herring season opened early. Ed and I scrambled, and when the opening came we were there: punts in the water, nets in good shape and so on. But we were not mentally ready.

I blame myself for this, as I had listened to a guy we call the Scientific Fisherman. We don't call him that to his face, but he got the name because he is always pointing out what a scientific fisherman should do in given circumstances. According to him, that particular year a scientific fisherman could relax, as the herring would be late—there was very cold water in the Gulf, rough weather, moon was in the apogee (or maybe it was the perigee, whatever). Unfortunately, the herring had not read the same books that the Scientific Fisherman had, so we were caught unprepared for the early opening. We cursed Scientific as we rushed over to Vancouver Island to go to our old fishing spot at Denman Island.

To make a long story short, we had good fishing although we had to put up with some very bad weather as well as the usual anxieties of herring fishing. But once back on the beach, we were able to put our minds into the springtime mode. So I got to thinking about spring fever. Not the birds and the bees and all that stuff, but what it is to be in the fishing racket and what we should be doing in that season. One aspect of the longer days of spring is more sunshine, which meant we would be painting boats and stuff.

The Scientific Fisherman said this was all wrong. Paint is to preserve things, so we should really paint in the fall, because winter has the lousiest weather. We all agreed with him, but since we were all getting ready to paint we just went ahead. At least thanks to Scientific—if you are willing to believe him—we found out porch paint is the same as marine paint, but a lot cheaper.

He always has an opinion on everything and has lots of ideas. His knowledge is like a cheap blanket, wide but thin, but at least he does have an inquiring mind and does not get his opinions from radio talk shows. He is a true believer in his opinions because he feels he has arrived at them through a scientific process, therefore they must be true.

Spring means people open their gear lockers and start talking to net salesmen. Spring means Scientific starts figuring out what is the smart thing to do this year. He noted with approval that the halibut fishermen were out early because

Spring Fever and the Scientific Fisherman

of the price they think they can get; this he felt was scientific. I know for a fact that he has never fished halibut, but let it pass. Spring means fishermen start doing their taxes and I start hearing from my old customers, as I have a little computer program that does it slick as a whistle. The computer, however, hasn't convinced the fishermen to change their habit of just putting all their receipts in a shoebox and dumping them at my place.

These are the usual humdrum things of spring, but lurking in the shadows at all times is the possibility of spring fever. I am afraid I had an attack, brought about by hanging around the Scientific Fisherman.

Here's what happened. Somehow or other, Scientific got tangled up with an American guy from Idaho. I thought Idaho was all potatoes and nothing else, but surprise—it has a fishing industry, for gosh sakes! On some of the lakes and rivers, they fish a bit commercially plus they have some sort of an aquaculture industry. The American guy was out here because he wanted information on fishing techniques and somehow he got referred to Scientific. Scientific called me because of my notebooks.

I went over and met with them and checked out Mr. Idaho Potato. A small, neat, wiry red-headed guy with piercing blue eyes and a quiet voice that could persuade any widow to cash in her savings bonds.

You guessed it, I had a sudden attack of spring fever. Two things were going on as we talked. On one level it was all scientific as we talked about the maximum pull of small power blocks, how deep you can fish with a seine, the efficiency of mono-filament gillnets and so on. On the other level he and I were checking each other out as the Scientific Fisherman rattled on. To make a long story short, Mr. Idaho and I went to Stanley Park, toured a couple of museums and, in the short time allotted to us, did pretty well all the spring fever things.

Scientific was aghast. Nothing to do with fishing, he said. At the very least, not very scientific. "Spring fever!" he snorted to me after Mr. Idaho left.

One man's science is another man's mumbo jumbo. Scientific, like all his type, is fascinated with gambling odds and feels he has a system that will make him a fortune. He said he was going to Las Vegas with his system and I said he wasn't scientific at all; he merely had a dose of spring fever like me.

He didn't even acknowledge my statement. That's another thing about spring fever—it affects your hearing.

Spring Cleaning

Spring cleaning comes later for fishermen than for others. One reason is the herring fishery, which takes place in March. Also, salmon season starts later and later each year, so one way or another it is not until late April or May that spring cleaning gets done on a fishing boat.

When I went into Steveston to get the stuff to clean Ed's boat, signs of spring were everywhere. The lineup at Sara's Ice Cream was full of tourists and the tables in the Net Shed were full of fishermen. I guess the gossip and BS-ing were increasing along with the daily pollen count. Fishermen have lots to talk about in ordinary times, but what with endangered salmon streams, short openings, tighter regulations that make you buy more gear, face-offs with the Americans and the whims of El Niño, these are not ordinary times. Everyone agrees that something should be done but no one knows what.

This is a good time to talk to my pal the Scientific Fisherman. Scientific is good to have around when you are doing all those millions of things that go under the heading "cleaning up the boat." There are many things you must know, and you must know why you're doing them.

When you go on the boat for the first time (I'm talking about Uncle Ed's troller), you first give it a look to be sure it has not sunk. The little automatic bilge pump has to be hooked up to shore power. My dad sunk a boat because the pump was hooked up to the engine battery. Dead battery, no pump, sunk boat.

If the boat is still floating, you go aboard. Then you try and find the key so you can go inside. Scientific claims that he can find the key to any boat in about five minutes, because the hiding places are so obvious.

Opening the door, the first thing you do is smell. I don't know why, but you do. The smell of a boat is a mix of diesel, mildew and fish.

Because of changes in temperature, condensation builds up inside, so to stop moisture from accumulating in all the little places it shouldn't, we put little heat lamps in the boat. They are just enough to keep the temperature steady, so even though these things are on, you want heat before you go to work, so the galley stove is your first target. Stoves can be tricky. They run on fuel, not on water, so they have a little filter and a "settler" that collects the water that may be in your fuel tanks and can build up in your lines. Once the stove is going, coffee is made and cleaning can start.

There is a big philosophical problem about spring cleaning on a boat. Besides cleaning, there are a million tasks that should be done at the same time. On a fish boat, everything relates to everything else. If you need to clean the rust off the top of the stove, shouldn't you also clean out the carburetor? One is cleaning, the other really is maintenance. And so it goes.

Spring Cleaning

Garbage is always a problem. Things accumulate on a boat and spring is the time to get rid of them. Out go the tide books from last year and the year before. Also the girlie mags that always appear but no one admits to buying. Old magazines. Ed is a confirmed reader of gardening books even though I have never seen him do any gardening. Strange. Then there are bits of gear that are useless, so they must go too. Every piece has to have a decision made about it. Sometimes you end up with a thing and you have no idea what it is. It is obviously a part of something, but what? And why was it saved? Throw it away and sure enough you will find it was an important part of the power steering or something.

Scientific always gives me and my pals a little lecture on how there should be a scientific program of maintenance and systems and maybe even computers for fishboats. Poor Scientific! I know he can't type, so getting a computer would be like trying to play the piano without a lesson. His deckhand keeps all the electronics in order. Scientific just runs the radar, because it only has two knobs and an on-off switch. So all his talk is just that.

Don't get me wrong, his talk is always interesting and sometimes useful. But spring cleaning is still spring cleaning. You need elbow grease and a determination to get it done. However, Scientific did give me a good tip about how on a boat it is different. When you get tired of it, you lock the boat, hide the key and go home. Out of sight, out of mind. "It will still be there when you come back," he says. And that's a scientific truth.

The Case of the Left-handed Halibut

"There is something wrong with that fish," said Uncle Ed.

My heart sank. I had just brought the halibut to Ed's house in Richmond from the dock where I had got it from Barry.

When Barry is around there is always trouble for one reason or another, but the fish as it lay there on the floor looked pretty good to me. It was a fresh halibut, only about four days old, that Barry had picked up as a crewman on a halibut boat. "Real stuck gear," he told me, "none of those damn snaps." He meant he was really an old-time professional, using traditional gear.

Right now that didn't matter. I had offered some fresh halibut to Uncle Ed and he had found something wrong with it. I took a second look and saw right away what he meant.

The Case of the Left-handed Halibut

The halibut was left-handed. That is, as it matured and went from a normal fish position and flopped over on its side (and one eye migrated over), this particular fish had gone a different way from the tens of millions of his species. The result was a left-handed halibut.

I know a little bit about halibut fishing—just enough to be dangerous. My husband and I one year had foolishly rigged our boat for halibut with snap gear—a longline where the hooks are attached with removable snaps. Foolish because we didn't really know how to fish halibut. All we had was desire and some marks on an old chart. Our plan was to fish halibut with the pros and then dazzle them with our catches in "soaker holes." The theory is that soakers—which are very large halibut—lie in holes in the bottom of the ocean. There is a lot of competition among the halibut for these holes, so as soon as you catch one, the next big soaker in the neighbourhood finds out about it and moves in. Find a good soaker hole and you will become rich, or so we thought. It was not to be. It is tough fishing. While we didn't starve, we were lucky to make minimum wages. So ended my halibut career.

But halibut is the finest kind of fish to eat, especially if it is fresh. Nowadays there are lots of fresh halibut around because the rules have changed. Under the old system, there was a "gold rush" system where all the boats fished like crazy to get their share before the quota was caught. This generated too many fish for the fresh market, so most of the halibut were frozen and the price fell accordingly. The present system gives each boat a quota which they can catch whenever they want. The result is a continual supply of fresh halibut, and the fishermen get a better price. The halibut I got from Barry was about 25 pounds and was worth about 75 bucks give or take a little.

Ed and I figured out what was wrong with the halibut, but we decided it didn't matter because we would cut it up anyway—some of it into steaks to potlatch away, some to eat over the next week or so, and a little to freeze.

In the end, Uncle Ed took over most of the cutting. He was in his element when he had a chore he liked, but was free to express how much he didn't like it and why. The whole time Ed was cutting up the halibut he swore and muttered things like "morphodite" and "gunnysack" which I assumed applied to the fish, and "red-headed Jasper" which I knew applied to Barry. He also lectured me on the merits of "God's will," which, loosely interpreted, is what Ed thinks it should be, and the necessity for right-handed halibut. More or less summarized, Ed pointed out that it is impossible to fish on a halibut boat if you are left-handed, and impossible to clean one with the knife in your left hand. I didn't argue even though much of what he said was plain wrong. I needed the halibut cut up and safely in Ed's freezer.

The other thing on my mind was Barry. Why had he given me the fish? There was no love lost between us. Barry had been the beachman on the little seiner my husband and I had. He drove me nuts because he was always causing trouble for no apparent reason. For example, once a big row broke out on our boat between the engineer and the cook. It had nothing to do with Barry, but Barry had caused it. He started with me. "You look troubled," he said. "Is something the matter?" I said no and thought that was the end. Not so. I overheard him ask the skiffman the same question. Then I heard him ask the engineer the same question. Having gained everyone's confidence, Barry then stirred up trouble by dropping bits of information about what one crew member had told him about another. The row between the cook and the engineer was the result of Barry's meddling.

That was Barry's style. Now he owned a troller and could bother whoever he wanted.

I mentioned all this to Uncle Ed. Ed declared it was all as plain as the nose on my face. Barry, he announced, was "one of a kind" and so was the left-handed halibut. It was God's will that people like Barry were produced, and so were left-handed halibut.

As usual, Ed's logic eluded me. Ed might not be logical but he sure can cook halibut. Left-handed or otherwise, it tasted great.

Eulachon Madness

The month of May has many faces, and fishermen know them all. We are close to the weather and are attuned to the many changes that late spring has to offer. The weather is ever on our lips and in our hearts.

But last May was one of disgusting rain. We know it is good, because it puts snow on the high mountains and fills the land with water for the summer. The returning salmon will find water in the streams. But we dwellers of the rain forest long for the rare pleasure of being able to complain about the heat.

May brings other things besides rain. It also brings the joys of early summer. One day last May something happened that took me back to my childhood. The phone rang and the speaker said crisply that eulachons could be had for the asking—but hurry because the word was out.

I dashed for the pickup with a handful of plastic bags and was off to the fish plant. One of the seine boats that was on a late charter on the herring had come down from upcoast. Somewhere, somehow they got several tons of eulachons.

Now eulachons are curious things. They are a type of smelt that are unique to the Pacific Northwest. Early BC settlers called them candlefish, because they contained so much oil you could light one up and burn it as a candle. The Indians used the oil the same way Italians use olive oil—for everything. But to me, eulachons mean only one thing—Digging Sam.

In the village I grew up in, Digging Sam was called that because that is what he did. He earned his living as a digger. Sam's specialty was digging of the plain variety. Ditches or basements or a potato patch were the sort of thing he was the master of. Like all small towns we had too many graveyards, and the soil in each was different, but Digging Sam knew the best way to approach each and all of them. He dug graves for a set fee, regardless of race, religion or location.

Beyond that, he had no talent. He couldn't get work as a rough carpenter, because he couldn't even pound a nail straight. Most labouring jobs were also beyond him. So with meagre skills, Sam specialized, as his like have always done, in those tasks that are so mind-numbing and hard that even fishermen will pay someone else to do them.

Digging Sam had a few vices and a few virtues. His vices were simple and consisted of a fondness for beer, but he only got drunk on Saturday night. He always showed up for church on Sunday, and would occasionally burst forth in song when a favourite hymn was played.

Overall, Sam's life had little meaning until eulachon madness struck. That's what he called it when the Indians would bring the eulachons to town from the Inlet. Sam loved eulachons. And in his mind it was clear that if he loved them—rolled in flour and fried in bacon grease—then everyone else must also love them.

He translated his love into action. He would get a wheelbarrow and load it up with eulachons and start on his journey through town to those who were to receive them.

That's how I first met Sam. Late at night there was a brisk pounding on the door and a voice called out, "It's eulachon madness again."

"It's Sam," said my dad. Out we went to the wheelbarrow and got a bucketful. My father thanked Sam profusely for the gift. "Don't thank me," said Sam. "It's madness." Without any more ado, off he went to his next call.

Why Sam did it, no one ever asked. It was just accepted that in May a knock would come at your door and there he would be—smiling, happy and in a rush. There is probably no significance to it all. It is merely one of those things in village life that just happen, and it was all when I lived on the Island with my dad, long ago.

When I got home with the eulachons, Ed was there waiting for me.

"Where've you been?" Ed asked.

"It's eulachon madness," I said and heaved a bag of eulachons into the back seat of his car.

The Great Eulachon Hunt

I was in Sara's Ice Cream shop in Steveston, getting some ice cream, when who should walk in but Uncle Ed and his pal Pete. I asked Ed if he wanted to buy some smelts from the fish store, as I had just bought some.

Ed snorted and said real men don't buy smelts, they buy eulachons. When Ed wants to get my goat he tells me what "real men" do or don't do. I snapped back that real men didn't buy eulachons, they went and caught them. "Issat so? Issat so?" said Ed. Little did I know that the Great Eulachon Hunt was now on.

Eulachons come every spring from the ocean to spawn in the major rivers of BC, and the Fraser River has a nice run. That was what Ed had in mind.

I knew he was up to something when I drove by his place and saw something green in his driveway. Being nosy, I went and had a look. It was the most bilious green twelve-foot aluminum skiff I had ever seen. Attached to it was a five-horse outboard. It was sitting on a little trailer and even had a sticker pasted on the bow that said "Guide tested—Guide approved." While I was looking, out the door came Ed. "What the hell you looking at?" he asked.

I remarked that it was a nice skiff and outboard, but was Ed going to get a trusty Indian guide to help him out? Ed snorted and said when fishing eulachons no guide was needed. His pal Pete was going to help. I didn't say anything but smiled to myself. The last time Ed and Pete went on one of these expeditions was moose hunting, and Pete ended up paying for a dead horse he shot in error. "Nice skiff, eh?" said Ed. "Got a deal, 200 bucks for it and one-fifty for the outboard." By my calculations Ed was into the deal for big bucks already, so he must be serious.

I was busting to find out how they would make out. Every time I went by, the green skiff was still sitting in Ed's driveway. Ed assured me that there was only one way to get eulachons. A few days before the full moon, on a flood tide just at dusk, in a secret spot in Ladner. On that appointed day, Ed noted, I'd better be standing by with as many buckets as I could rustle up, because by about nine-thirty in the evening he would have enough eulachons to feed an army.

One day the ugly green skiff was gone and I knew the hunt was on. I waited for the skiff to return but it never showed.

Finally I rang the doorbell and went into Ed's place. "Where's my eulachons?" I asked.

Ed, as usual, was sitting at the kitchen table drinking instant coffee. He looked up at me pityingly from his newspaper. "You snooze, you lose," said Ed.

According to him, the demand for his eulachons was so great that he had none left for me. Something was wrong but I didn't know what.

The next week I was in Ladner on the dyke in one of the net lofts and I happened to look downriver. I saw something green—in fact, it looked like the distinctive green of Ed's skiff. Out of curiosity, I walked along the dyke and went down to the little float, and there was the skiff and the outboard with a beat-up smelt net in it. Two men approached me, as I was obviously trespassing on their float.

"I'd like to buy that skiff," I said. It was not for sale, they said, and a long silence ensued. They asked why I wanted it, and I replied that it looked ideal for use in catching eulachons. They both burst out laughing.

It didn't take long to get the full story. The week before, a friend had phoned them to say that it looked like someone was trying to steal stuff off their float. They hustled right down and found the skiff, with Ed and Pete in it, tied up at their float. The net was caught in the outboard's wheel—in layman's terms, the propeller—so they had drifted downriver and tied up at the float. Pete had fallen overboard, of course, and was soaking wet. Ed was swearing and hacking away at the net wound up in the wheel.

After some discussion with the men, Ed and Pete asked for a ride back to their car. "What about the skiff and motor?" asked one of the men. "Gimme 50 bucks and a ride and she's all yours," said Ed. And so it was done.

So the Great Eulachon Hunt was over. Ed was out several hundred bucks, I had to buy some in New Westminster to get a feed, and the eulachons—well, they are in no danger of becoming extinct.

A Piece of Rope

When my ex-husband and I fished halibut with our first little boat, we used "snap gear" and I developed muscles in my thumbs. You push in each snap with your thumb and attach it to the longline. The snap has a hook attached to it which you bait. But last season, the halibut fleet went out without me. I didn't have what halibut men call a "chance." Through some twist of language a "chance on the halibut" means the opposite. It means you have a firm offer of a job. Anyway, I didn't go out. I'm scheming to get a job either on a troller or a seiner, but without a family connection it will be tough.

I felt kind of low about it but brightened up when my friend Mary phoned. She wanted me to help her get some rope. I smiled when she used the word "rope" because people in the fishing industry never use that word. They say "line."

I went over to Mary's house to pick her up. She lives in a strange place by the airport called Burkeville, which has a hundred or so houses built during World War Two to house aircraft workers. They are the darnedest looking things because they are very small and were built cheaply and in a hurry, probably out of two-by-fours and shiplap.

All the fixer-uppers must have moved to Burkeville, because there always seems to be construction going on. Because the houses are so small there is no storage space, so all the building materials are stacked in the yard or propped up against the house. By glancing at the yard you can tell the owner's next project. The houses start to look like decorator crabs, which pick up whatever is in reach and attach it to their shells.

I like Mary because she suffers from the same problem that most women in the fishing industry have. That is, no matter what she has to say, no one listens or appears to take her very seriously. Mary and her friends are hikers and rock climbers. She is very good at it but she complained to me that no one listens to her ideas. I tell her that she might as well be fishing.

I remember one time we were tied up in Ucluelet and a bunch of us trollers were talking about knots. Fishermen have a bunch of different knots for their gear. After listening for a while I offered some information I had about the use of the square knot on nylon line. I pointed out that it "capsized" and came undone on occasion. I then showed them a better, special new knot, called the Hunter knot, after the inventor. I was pleased with myself and expected everyone to be impressed. All I got for my trouble was a strained silence. Women aren't supposed to know more about such things than men, I guess.

So Mary and I went to Steveston to get some rope she needed for hiking on the West Coast Trail, which runs from Bamfield to Port Renfrew. Her husband didn't want to take their special Swiss-made climbing rope that costs a dollar a foot and looks like it was made from Jacob's coat of many colours. So I gave her my lecture on rope.

In the old days fishermen used manila line that had to be stretched if it was to be used for cork lines or a purse line. Some was three-strand, some was four-strand and so on. The new lines are made of synthetic materials. Generally polyprop is cheaper and it acts more or less like the old manila stuff. Unfortunately it has a number of bad habits. For example, sunlight ruins it, and when it breaks under a strain it is murder. It whips around like a snake. Tie-up men on seine boats sometimes ask for danger pay. The other kind of line is made from braided nylon, usually with a three-strand core. Upon these basic types there are a number of variations.

I explained all this to Mary in the stores in Steveston, at the same time keeping a weather eye open for anyone I knew who might have information on jobs.

It took a long time to go into the suppliers and check out all the various kinds of lines available and get what Mary wanted. Also, she showed me a lot of knots used by climbers, and between the two of us I guess we kind of rewrote the book on lines and knots.

Finally, as a treat for her I showed her two new lines. One is the stuff seiners use for their running lines—the floating lines that are let out as the net is set and then winched back in. Strange stuff, but it is skookum. The other was the new Kevlar stuff that costs about 20 dollars a pound, floats, and is stronger than steel. Mary couldn't afford, it but if I was a climber that would be the stuff for me.

What with visiting all the suppliers, practising knots and pricing, we used up the morning, so we stopped for lunch in one of the new little restaurants that are springing up as the yuppification of Steveston continues. When we got back to Burkeville we were exhausted from our research into lines and the best knots. We were happy about it and then sad. It would be no use for Mary to reveal her knowledge to her husband and his pals, and I was in the same fix. It's awful to know something and no one listens. All I wanted was a job, and all Mary wanted was a piece of rope, and the next thing we knew we were caught up in technological changes in fishing and in the problems women sometimes face in competing with men. Well, at least Mary got her piece of rope.

How to Buy a Boat

Since hockey season is still on, I decided to go out and buy a boat. Too many of my pals are into staring at the old boob tube. I point out that it is the month of May—not the time for hockey, but the time for getting up a group and bringing back Maypole dancing. "I always wanted to be the Queen of the May," I tell them, but sports freaks have no sense of humour and don't listen to me when I tease them.

So out of boredom I went to buy a boat, not just any old boat but a licensed fishboat for use in the Pacific Ocean. The boat was not for me, it was for Uncle Ed's former deckhand Frankie, the one they called the Non-talking Deckhand because that is what he did, or rather didn't do.

I went along to see just how you go about buying a boat in this day and age. Every fisherman thinks he knows how to buy a boat and what's more can tell good ones from bad ones. I was intrigued as to how the Non-talking Deckhand would do it—not that he didn't talk, he did. He explained to me that around Ed he didn't get much chance to talk, and if he did Ed would only argue with him. As Ed was his boss, he decided not to say anything.

So I knew he could talk, but I soon found out that he could also drive you nuts. He never used a sentence if one word would do, but the thing that got me was that he had the habit of always answering a question with a question. When I got fed up and asked him why he did it, he said, "Do I do that?"

None of all this was going to stand in my way as we went about buying a boat. We went to the Pond, over to the Annieville Slough, the North Shore, False Creek and even up the valley. Frankie was looking for a particular boat with two licences: one for shrimp and one for salmon trolling. "Looking for a married A-S plastic," he would say. In the jargon that meant a licensed fibreglass boat. Buying a boat is bad enough, but you have to be a Philadelphia lawyer to figure licensing.

It was fun tagging along and I made myself useful writing down all the details about each boat as we looked at it. Some boats we had photos of, others not. Some boats we liked, but not the owners and sometimes it was the other way around. Some boats came with gear, some without. Some had an engine log carefully maintained with a computerized analysis of the oil, which tells you the same thing a urine sample does for humans. Age of boat, price, size of boat, electronics and so forth. Every boat was guaranteed to make Frankie a millionaire, and the only reason for the sale was that the owner had too much money and was retiring to Palm Springs.

To tell the truth I didn't care what boat Frankie bought and I certainly would never guess what he had in mind other than "A-S glass." From the "glass" part I knew he wanted a low-maintenence boat, but you pay for that. The "A-S" part meant that Frankie was tired of short salmon seasons and wanted another fishery. But what boat would he buy?

I am sure the sellers never had an inkling of what he thought because he told them all the same thing: "If I'm real interested I'll be back." When pressed by anxious sellers he merely repeated himself. Now and then a laconic phrase such as "nice hatch" or "okay electronics," but never a full-blown analytical sentence describing his criteria for passing judgement. Three days of this and I was antsy. I was ready to quit when he phoned to say he was ready to purchase and would I help him with the offer to the broker. No big deal there. Frankie offered cash for the boat and licences validated by the Department of Fisheries.

In the broker's office, I suddenly started paying attention. The boat Frankie was intending to buy was way out of his price range and he was putting in a disgustingly low offer. Further, we had hardly looked at the boat. True, it was "A-S glass," but what was going on?

The next thing I knew, Frankie got the boat at his price.

I couldn't figure it all out. For helping, I got to use Frankie's pickup for a few days to move some stuff around, but I was still puzzled by the whole affair. Nothing for it but to ask Uncle Ed.

Ed was his usual charming self. "You don't know a damn thing about buying a boat," he said. According to Ed, Frankie set everyone up by being uninterested, making everyone think there were lots of boats for sale and few buyers. This I guess was true. But why did he get the boat at such a low price without even bargaining?

"Easy," said Ed. "Frankie don't talk, but money sure the hell does." Live and learn, I guess. Now I know how to buy a boat.

Down and Out on Minstrel Island

All fishermen suffer from the urge to try new fisheries. I guess that is because fishing requires a person who likes adventure of one sort or another. Sometimes when you want to try something new the best thing to do is lie down until the feeling goes away. When Ed's former Non-talking Deckhand Frankie asked me to give him a hand on a few trips while he took his new boat out shrimp fishing before the salmon started, I jumped at the chance.

Shrimp fishing on Frank's boat was to be my life for several weeks, and as we left False Creek I noticed how small we were compared to the big seiners I had fished salmon on. There were only three of us on board and I didn't have to cook but I am always willing to spell off the cook. When my former husband and I first got our little seiner I hated cooking because I had to do it.

The other thing I noticed is how slow we were. We had lots of fuel aboard and extra gear and we barely made nine knots. And bucking any sort of tide was pathetic. We examined the tide book in exquisite detail because we wanted to hit the Yuculta Rapids at the right time. Seiners chug right through the Narrows when the tide is close to what they want, but we had to be bang on and a longer way around. *Captain Lillie's Coast Guide* is your Bible when you are slow and small and travelling the coast of BC.

A shrimper has that big stainless steel beam that rests atop the cabin from the stern and lets you know what the boat is fishing. Ours was 32 feet long and as the boat was 45 feet you noticed that beam all right. The beam holds the trawl open as it skids across the bottom on its little runners.

No need for trawl boards and lots of power under our system. We tried out some spots that I had heard about and some that Frank had heard about. We got shrimp almost everywhere but that is not the point. You have to get them in big enough quantities to make money and they should be big as possible and you should have a good market. Our plan was to freeze the first catches we got and then sell the last part we got into a fresh market as the price would be better. The damn freezer ran off the main engine so it was going all the time. I learned to hate the noise. But worse was to come. The engine started to heat up on long tows, so we knew we had a problem. We needed a land phone to talk to a mechanic, so into Minstrel Island we went.

Minstrel Island is one of the nicest spots on the coast. Nice restaurant, liquor store, fuel dock and so on and it has a long history that adds to its character. But we were not able to fish and things looked pretty bleak. We read the engine log, phoned the mechanic, ran the engine for hours and plotted the temperature. The possibilities for problems were explored. Several schools of thought flourished. I had my money on the impeller blades in the water pump as being defective and worn. Frankie subscribed to the thermostat-acting-up philosophy. The deckhand swore by the faulty-temperature-gauge ideology. The consultant mechanic, 150 miles away, had no opinion. No reliable power, no fishing, no money. We were down and out on Minstrel Island.

I didn't really mind as I was only along for the ride and a few bucks if we got good fishing. The deckhand said nothing, as it was not his style to worry. Deckhands are like that. Frankie the skipper said nothing except to note that it was a fixable problem.

Frankie told me about the first time he came to Minstrel. He was a kid and he was hunting with his relatives from Sointula up at the head of Knight Inlet. They stopped in Minstrel for "sea stores," which means more beer and wine. Tied to the dock was a boat called the *Bear Point*. It was a crabber and it had a sign up that said "Crabs for Sale—Large $2.00 each, small $1.00" (remember this was some time ago).

Now Frankie laid the puzzle on me. How many large crabs did he sell, and how many small crabs did he sell? I said that there was no answer, because we didn't know where he was fishing, what were large and what were small and so on. The deckhand agreed. Frankie smiled and said there was an answer, because he asked the crab fishermen about that very thing and got the solution. To tease us, Frankie kept the answer from us as we waited for the diagnosis of our engine problems and the necessary repairs. The crab thing bothered me so much I forgot about the engine. There had to be a simple answer or a clue but I couldn't solve it.

The plane came with a new thermostat, we took out the old one, fired up the engine and ran it under a load (tied to the dock). It didn't heat up and so we were mobile again. As we were going by Protection Point, Frankie said to me, "He sold *all* the crabs as large, because every customer wants large, so the first person takes the biggest ones he sees, the second person takes the next biggest and so forth. The crabs get smaller, but to the customer the biggest ones he sees are always large." We may have been down and out but at least we learned something new. Maybe next time instead of going shrimping I'll go crabbing. At least I'll always get large.

No Cure, No Pay

Fishermen read anything and everything but I doubt if one in ten has ever read a Lloyd's Open Form Salvage Contract. I thought about this as I waited in the doctor's office for a cure for my saltwater boils. Like a damn fool I got a bad case of them on my wrist. I know better, and somewhere I have real "wristers" like halibut men wear—you know those woollen things that are just the ticket for stopping saltwater boils. But I went shrimp fishing up Knight Inlet and didn't bring the wristers, and had nothing to stop the sleeve of my raincoat from rubbing me raw. We made money, but I don't need these boils.

Lots of things go through your mind in a doctor's office and the salvage contract seemed to have some meaning for the doctor and me. We pay doctors up front through our health plan, but I remembered a Chinese custom of not paying doctors when you are sick but paying them when you are well. No cure, no pay. The same thing applies to the salvage of vessels. It works like this: if you are on a rock and sinking and the barometer and tide are falling, your rescuer hollers at you, "Will you take Lloyd's?" and you obviously shout "Yes." If you are rescued there are rules as to how much salvage your rescuer gets. I have never seen the Lloyd's form or been involved in the situation, but I sure like the principle of rewards based upon peformance or, as it is called, "No cure, no pay."

Fishermen understand the principle—as a matter of fact, they live by it. No fish, no pay. That's why one of the hardest questions fishermen have to answer is the one they always get asked: How much do you make in an average year?

There certainly isn't a short answer. Trying to explain the share system on a seine boat is an exercise in long division, percentages, history, fish prices and salmon cycles. After two minutes your audience is completely lost. The reason is that most people are conditioned to sure things, such as a wage or a salary, and in most jobs the link between performance and pay is very poor. In fishing it is direct and simple.

So here's my plan. Fishermen can start asking others to live by the no-cure, no-pay idea. Think what it could do for the health care system and for the schools, never mind your car mechanic. Of course it will never happen, but I am sure this coming season many fishermen have dark thoughts about how things are going and want a good number of changes. Salmon fishermen are beside themselves with the prices for fish, the treaty with the US, fishing time and so on. It looks like a tough and frustrating year. For starters, the Department of Fisheries and Oceans will have to bring exactly the right amount of fish at the right time and place or no salaries for them. Second, the treaty with the US has to be settled to our satisfaction or no pay for the boys from Ottawa.

No Cure, No Pay

I am sure that everyone can apply the principle in lots of other ways. But like all good ideas, it is doomed to failure because one person's great idea is another's crackpot scheme. The fishing industry is loaded with great ideas I am sure, and in these troublous times we sure need them. That's why we need first principles like no cure, no pay.

My accountant's first principle is: get next to the money and all is solved. Uncle Ed's is simple: do it now. My ex-husband's was: delay doing anything, things change. (You can see why it didn't work out between him and me.) One of my favourites is that of the Jesuits: never ask for permission to do something, only ask for forgiveness.

You can see that hanging around a doctor's office can lead to poor mental health, but saltwater boils are no fun, especially when I got them by violating one of my own first principles: don't do anything dumb. And it was dumb not to wear wristers. But I ran up against another first principle: KISS, or keep it simple, stupid. Because I'm on a small shrimper, I am trying to keep my gear and stuff to a minimum because space is always a problem, and being a woman on a small boat I am extra cautious about how I do things. So I didn't pack my wristers, which I have packed on and off boats since my husband and I fished halibut on our first boat so many years ago.

I got the salve from the drug store and some pills just in case, but prevention is the best medicine of all. I am trying for a job on the salmon and it looks like I have one on the west coast at San Juan when it opens, if ever. In the meantime, back to first principles: a positive cash flow is the secret of happiness. So I'm out here shrimping. No cure, no pay: no fish, no pay. Even I understand that.

The Old Captain Takes a Walk in the Sunshine

I watched him as he walked along the path to the parking lot at the Steveston fishermen's dock. I knew he was making his way there when I had spotted him earlier, over at the other dock where they sell fish and shrimps to tourists for cash.

I also knew who he was. He was an old fishing captain taking a walk in the spring sunshine. But it was not just any old walk. No, he was doing something that comes to us all: satisfying the need to relive the past by looking at familiar and unfamiliar things. He walked to look at all the boats tied up in Steveston, and to let his memory do its work. Sort of like reading a book many years later and finding things you missed the first time around, as well as the old familiar things that you were looking for.

I had time to examine him closely as he walked to the

The Old Captain Takes a Walk in the Sunshine

dock. He wore a black suit of heavy cloth that looked like it had been bought long ago, to be worn only on serious occasions. His shirt was white and his tie was so old it was about to become fashionable again. He wore sensible black shoes and though I couldn't see from where I was, I knew the socks were hand knitted, and probably hand mended. I was sure a wife or someone from the old days had made them. I even knew the label of the hat he wore. It was a brown fedora that went out of style with gangster movies, but it was not just any old hat. No, it was a Borsalino, and to those who knew it, it told a story. The old captain was presenting himself to the world and he had to do it just right.

The hat was the key, for I knew him even though he did not know me. When I was a kid on the Island, this man was feared and respected. He was a top fisherman, what we call a "highliner." The hat was given to him by a fishing company at the annual Christmas party, recognizing him as the top seine boat captain on the Island. They don't do things like that any more. Anyway, you can't wear a fedora to Hawaii or Palm Springs.

As the old captain walked by me I tried to figure out his age. He was over seventy and life had not been kind to his body. Arthritis in his hands from longlining, bad knees, and probably high blood pressure from the fisherman's diet of cigarettes and coffee.

But his eyes were not old. He seemed to have two ways of looking. One was the normal way, the other was what I call the zoom mode, where he could be looking at you and talking, but his eyes were scanning the horizon and checking every detail of boat rigging and how the water was and the state of the tide. A million details in a glance—that came from fishing in the days before radar and sounders.

Notes from the Netshed

Now he was down on a float, looking at the boats tied there. The aluminum super skiffs he barely acknowledged. Nothing like them when he fished. Same with the fibreglass gillnetters. But when he came to an old wooden boat, he paused and took a long hard look.

I am sure he knew many of the boats well. One thing about the BC fishery is that its boats last a long time—some more than fifty years. So the old captain could look at a boat and think of many things. The introduction of hydraulics that came with the Puretic power block for hauling in the net. The transition to drum seining with running lines, Gearmatics and so on. Nylon replacing manila and cotton, forcing fishermen to learn new knots. But it was always the fish that he came back to. Now I don't really know that, but I was trying to put myself in the old captain's place. I mean, I could have gone over and talked to him and asked him point-blank what he was thinking about, but I didn't. Somehow I thought it was private.

But what fish he must have seen. The pink run in 1962, when the fish came to the Namu area for months. The '58 sockeye year, when they came through the straits. And dog-salmon fishing for weeks on end. A season that lasted from June to November. Fish and places we would never know. They didn't have the power and the gear we have now. Sets were planned and carefully executed, not just banged out like now. Of course, the good old days usually weren't. No Workers' Compensation, no Unemployment Insurance and no pension plan. Prices were so low that fish were counted as pieces, not by weight like today.

My good old days are right now, and I want to make the most of them. That is why I watched the old captain with such care. One day I will be like him. I will come to the Steveston dock and look at the boats and dream of fish and places. I will look for my past in the boats, for fishermen grow nothing, build nothing, leave nothing behind. All we can do is look at a boat and capture what it was like for us. All of us have to take a walk in the sunshine; we can only hope that it will be a pleasant one.

SUMMER

INTRODUCTION

The Mystery Tour

One day the boats are all gone. In Steveston the "pond" where the smaller boats tie up for the winter is suddenly almost empty. The stay-behinds are the trollers that will only fish the Gulf of Georgia, and the odd gillnetter that only fishes in the Fraser River, either from habit or from just plain being tired of travelling the coast.

It's the same over at the government fishboat float. The big seiners are gone, leaving behind a few forlorn-looking derelicts with For Sale signs in their pilothouse windows. Up and down the coast, boats have left for the fishing grounds as it is summer, the time for intense fishing.

The return of the boats to the fishing grounds is tied to

the return of the salmon. They have their own clock and internal navigation system, and they know where they have to be and when to go there to spawn. Fishermen have to anticipate the fishes' movements and be at their favourite spots to catch them. It is not simple. Sockeye salmon come every four years, pinks ("humpies") have a two-year cycle, coho and chum salmon sometimes have five-year cycles, and the runs fluctuate in size. It is always a guessing game, so fishermen can only make their best guess and be prepared for some confusing things to happen. Some boats will leave Prince Rupert for the west coast of Vancouver Island and some will leave Vancouver Island for the grounds near Prince Rupert. It makes sense to a fisherman. If you are an outside troller you will go to the west coast no matter where you tie up your boat. The gillnetters start in the north and work their way south as do the big seiners. Each has its own pattern.

Some boats end up fishing the whole coast over the summer. In the 1980s we trolled for spring salmon off the west coast of Vancouver Island, tried fishing halibut in Hecate Strait, trolled again off the Queen Charlotte Islands, and later put on a gillnet for fishing chums in Johnstone Strait. We were obviously a combination boat. A summer of fishing was a tour of the coast, although we were not tourists but dead serious about making money. However, it was a great life. Beautiful scenery, no civilization, the joy that comes from doing what you want to do, and nothing to worry about except the fish. Life on the water has its special benefits. We saw whales, birds, seals and sea lions, and we could get our favourite food from the sea. Like fine restaurants, each place has its specialty. For oysters you must stay in the Gulf of Georgia, as they only live there. Clams are everywhere, and so are mussels, although I seemed to be the only one who would ever eat them. For halibut we all had our favourite "soaker holes" where the big ones were, and any reef or rock pile would do for jigging rockfish. Sometimes you'd get a surprise—once jigging for rock cod I got an octopus. Preparing, cooking and eating it was an all-day task.

While we fish the entire coast, some places are special. Everyone has their favourite and mine is the little town of Bamfield. Getting to your favourite place can be relatively easy, or as complicated as you want to make it. In general, fishermen are not big on navigation. To get to where you want to fish, following someone else is the easiest way and many boats do just that. Others go from memory, and others do it the textbook way. Using charts or guides that lay out the course is the formal method. Fishermen are lucky in that since George Vancouver's time the coast has been surveyed and marked in exquisite detail, and the coast is dotted with lighthouses, beacons, buoys and fishing boundary markers. It is possible to go from Vancouver to Prince Rupert in complete darkness, following the courses in a book from lighthouse to lighthouse. No matter where you are, there is almost always a blinking light off in the distance. In fact, each light is timed to blink at a different speed, so with a stopwatch you can determine exactly which light is which to figure out where you are—although I've never heard of anyone actually doing that. In addition, all boats now have sounders, so they always know how much water they have under them—a very useful piece of information. So if you see a fishing boat, you can be fairly sure the skipper knows exactly where he is and where he is going. To the despair of the bureaucrats, fishermen don't take much to formal training in navigation. Their needs are simple: getting to the fishing grounds as quickly as possible.

Once on the grounds the dream of us all is to be "on fish." Being "on fish" means that you have figured out some sort of cycle of the movement of the fish through all stages of the tide and you have got yourself into the loop. It's a nice feeling.

On the other hand, the anxiety of not knowing where the fish are is terrible. Sometimes you can be desperate for information. But information always costs you something. You have to give to get, and if you are on fish you usually don't want to tell anyone except those you owe information to, so fishermen have their secret radio channels over which they share this information. The whole business has become very high-tech, with scanners and scramblers, so if you look in the pilothouse of a large seiner you will be aghast at the number of radios—ten is considered an okay number with a minimum of about five. Every skipper has a little list of these radio frequencies and who can be found there, so eavesdropping is a popular pilothouse pastime.

Fishermen use a variety of gear to catch fish. In the salmon fleet there are only three: gillnetters, seiners and trollers. I have done them all. Trollers are hook and line fishermen who "troll" their gear through the water. It is my favourite way to catch fish because it is (or was) a very independent way of life. Trolling seems to attract people who like to do things in their own way and in their own time. Because they catch fish one at a time, I believe they are closest to the essence of fishing. When you're trolling, there is just you in the cockpit in the stern of the boat, trying to figure out what is happening several fathoms under the water with all the fish. If that doesn't make you deep and mystical, I don't know what will.

I tried gillnetting with my husband and the truth is I didn't like it. Most of the fishing takes place at night in the dark. You put your net out and let the tide carry it and wait for the fish to be caught. It made me fearful. Every little noise or creak of the boat signalled a disaster. Any far-off light could turn out to be a tug with a large tow such as a log barge that could run over your net and tear it to shreds. In my case my apprehension was probably linked to lack of sleep. I just couldn't train my body to sleep in the day. Also, the fact that we never caught many fish may have had a lot to do with it.

Now I work on a salmon seiner whenever possible. Jobs are hard to get, but I am getting older and a skipper and crew of five or six (that extra one is me!) on a seiner means extra comfort. Wheel watches are easy with two people, and there is a designated cook and an engineer, so those departments need not concern me. The rest of the crew has the responsibility of getting the net out and getting it back. My job is winding the net on the drum. It sounds easy and it is simple, but it's also fraught with peril. It has to be done just right, so that there is no dreaded backlash where part of the net, instead of going out into the water, perversely gets wound back onto the drum. Then there is the dreaded roll-up, where the lead line somehow or other takes it into its mind to start climbing up the web and rolling it all up into a tight black coil several fathoms long, ruining any chance of having a fishable net.

Writing about fishing is sort of like dancing to describe architecture. There are endless technical matters and tricks of the trade, but no matter how you fish there is always the indescribable joy of it all. Okay, it can be miserable: no fish, a cranky skipper, a sullen crew and worse. Like being in a floating jail with drowning your only way out.

But summer has sunshine and at least one day in every summer it is all perfect: brilliant sunshine, a little westerly breeze to cool you off, lots of fish in the hold and the skipper smiling. You are on fish and all's right with the world.

That's what brings me back every summer. I need the rituals, the simple routines, the patterns of fishing and the familiar coastline and anchorages. In the end, fishing is about people. The community I know is strange, but we know that we are all in it together. Like passengers on a mystery tour we assemble each summer, we go somewhere with a common goal in mind, and when summer is over we all go back to where we came from.

But we have had more than a journey, we had sadness and fun and ultimately some joy in a life that is different—and, sadly, is vanishing. But the fishing life has given me a reason to eagerly wait for summer next year.

The Sunshiners

Ah, the sunshiners, the sunshiners, the sunshiners. God, how I love the sunshiners.

You might ask, who are the sunshiners? They are fishermen, but fishermen of a special kind, and some of them came to visit me. This is why I'm telling you about them.

Fishermen come in all kinds and sizes and types. BC fishermen certainly do. If you go to Ladner you meet one kind, in Steveston another, in Rupert it's the Norwegians, and so on.

At one time, certain kinds of fishermen specialized in a particular fishery or area. For instance, the Norwegians dominated longline fishing for many years, and were prominent in the halibut fishery. The Yugoslavs brought the seining technique to BC, and the Japanese went into gillnetting and trolling.

As a new fishery developed or a new area was opened up, it attracted those fishermen who were most suited to it. So an old fishery like the halibut is still dominated by the

Scandinavians, and a new one like the herring roe is made up of every type from all the other fisheries.

Now for the sunshiners. The term was at first an insult, and maybe it still is (the worst insult I know of is to call a fisherman a "farmer"). Sunshiners were fishermen who only fished in good weather.

BC is a temperate rain zone. Sunshiny days are the exception, in spite of what BC-boosters tell you. So a sunshiner is someone who would not be found fishing very often.

Don't expect to hear from them early in the morning, or to find them travelling the coast at night. Heavens no, it's dark and you would have to use a compass and the charts.

However, sunshiners do like to travel. They are always looking for the mother lode of fish, so they will leave good fishing to go to another area—maybe there's even more fish there, and more sunshine.

From what I have told you, sunshiners must sound pretty bad. But times have changed and so have the sunshiners. The old type of sunshiners still exist, but they are getting scarce. Fishing is now a tough business, and they couldn't last. Some still survive, and their habits are the same. But my own definition of the sunshiners has changed, because the sunshiners I know have changed.

The general type of sunshiners is one thing, but the specific ones are friends of mine. In fact, they are somewhat distant relatives, so they come to visit from time to time. Being sunshiners, they never phone ahead to make an appointment. No, they just drop in after a three- or four-year absence and ask "how's it going?" All the doom and gloom in the industry vanishes as soon as they show up, because—well, if there's rainy weather, you move on and look for sunshine. The sunshiners avoid gloomy people like the plague, so they are not ones to find things too terrible in the fishing industry. They have always found sunshine before, haven't they?

The sunshiners, of course, make Uncle Ed mad as hell. Ed is a compulsive worker. "A driver," as his Non-talking Deckhand used to describe him. So the sunshiners and Ed are merely civil to each other.

Their boats are certainly different. Ed's is neat as a pin, with everything perfectly maintained. A sunshiner's boat is a disgrace to the tourist board.

My relatives tie up in Victoria, and to look at their boat is an education. On top of the cabin is a beat-up old Coke cooler, now being reused and recycled as a meat cooler. Next to it is a pile of old lines of every variety that, who knows, just may come in handy someday. Behind the boat, the seine skiff is riding mighty low in the water. That's because sunshiners like to explore, so they have cut a big chunk out of the transom to accommodate an outboard motor. Or, if it hasn't actually been sunk yet, it's up on the boat, with the outboard motor tied to something nearby.

In the rigging are old worn-out flags of companies the sunshiner may or may not have fished for, and every corner of the deck is cluttered with used parts, worn-out hydraulic fittings and leaky hoses. And that's just the stuff you can see up on deck. If you went in the galley it's more of the same, except for the Playboy calendars.

While Ed is fundamentally in conflict with the sunshiners, I love them. They are survivors. It took me a long time to figure it out, but they are right. Rather than fight the world and nature like Ed does, they go with it. They take what they want and if it isn't there, they move on. They look for the sunshine. If there are no fish, move to another place. Or get into the skiff and go dig some clams. Or jig some cod. Things don't have to be perfect, and you don't have to be forever maintaining things. No, just do what you want, and sooner or later there will be a sunshiny day.

The sunshiners cheer me up. The worst salmon season in years might be coming up, and would the sunshiners be worried? Not a bit. You see, they forgot to buy a big new boat and weren't too good at getting bank financing. So, they are stuck with their old junky boat—and no mortgage.

The value of their boat has doubled and then halved in a few years, but this is of no matter to the sunshiners. To sell the boat, they would have to fix it up, and to fix it up—well look, now it's started raining, and everyone knows you can't work in the rain. We've got to keep an eye on the sunshiners. They may not have the secret of the universe, but they've sure got the secret of something.

Working for the Yankee Dollar — We Build a Net

To fish you need a net and to have a net you must buy one or make one.

Every morning lately I have been driving across the Canada–US border to an industrial park in Bellingham, Washington where we are building a salmon seine from scratch. Through a network I didn't even know existed, I got recommended to this American guy to help him build a salmon seine for a Bristol Bay gillnetter who bought a "stateside" seine boat and somehow or other didn't want to or forgot to buy the net. So he needs to get one.

The Yankee guy decided to build one from scratch and needed someone to tell him how to do it. All he wanted from me was the little diagram or table of numbers that gives you the percentage of web that hangs where, and the number of corks per fathom, and lead line weight and so on.

There is no book you can get from the library that tells you all this stuff. Usually the skipper or one of the crew carries around all the details in his head. I, being me, have written it all down and I was only too happy to sell my knowledge to the Yankee imperialists. Well, actually his name is Dwayne. He pays cash on Fridays and those Yankee dollars are big ones.

In reality, building a seine is no big deal except that hardly anyone ever does it. Once built, a seine is continually upgraded. Old or damaged web is replaced, a new lead line may be added one year or a new cork line another year (it's not really a cork line since cork floats have been replaced by plastic ones, but it's still called a "cork line"). Special care is taken for the "money bag" or "bunt," because that is where your fish finally collect.

Also, the government now has regulations about the size of the mesh in the bunt, so there is an annual inspection to make sure you haven't shrunk below the limit. A million and one details, all of which are usually left to one person's memory.

To build a seine from scratch you have to have the overall design thought out. The net is designed for a boat and, to some degree, for a fishery.

A small boat with not much power means a light net with a light lead line. A light lead line means small corks (plastic floats). The big question is how much web to "hang in" to these lines. This is referred to as "percentage." There are other details, such as the tapers at the end of the net where the depth of the web is shortened, and the length of the line from the cork line to the lead line, which is called the "gable end." These are important details if you're going to make a net that is fishable.

The overall design has evolved by that method so

familiar to the fisheries—trial and error. You hear all sorts of stuff such as "You are really towing on your lead line" or "You should put more percentage in the middle" or "Heavier lead lines sink faster." All this is well meant and important, but no one really knows what the net looks like under water as it fishes. This doesn't stop anyone from being an expert on the subject. Like Uncle Ed, they are often in error but never in doubt. When it comes to building a net, no one needs to worry about being wrong, because everyone knows there is no way to prove any of the theories.

The practicalities of building a seine from scratch are simple. You need a general idea of your design and at least sixty feet of smooth floor surface. This is because the net is built in ten-fathom stretches (ten fathoms equals sixty feet). The fathoms are usually marked off on the floor and other marks are made for the various added percentages of web you will hang into the ten fathoms of cork line. There are other marks for the lead line as it is hung in shorter than the cork line. It was these marks and percentages that I was being paid for.

The Yankee wanted a net that fished and was bringing Canadian technology to the good old USA. The American had never fished a drum seiner so he wanted to get this part right.

Well, we did get it right. It was a strange crew that built the net, though. Me, two Mexicans, two Norwegians and an old Slav guy. That's how it is in the fishing racket—all kinds of people turn up and the only thing that matters is, can you do the job?

The Mexicans taught me a new word. The ring fitting at the tow end they called an "oertza," which I am going to look up in a Spanish dictionary one of these days. The Slav guy didn't tell me anything other than to say he was a

Yugoslav, and the intensity with which he said it meant I didn't ask from what part or anything.

The Norwegians told me a joke that only a few of you will get. They were from Ballard, the part of Seattle where the Norwegians settled. Question: How much does a haircut cost in Ballard? Answer: Four dollars, one dollar per side. As far as I was concerned, getting the joke was optional, as long as I could keep getting these Yankee dollars.

Two Matches and an Attitude

You can't tell a fisherman by looking at him or her, as they dress more or less like everyone else. And if you look at a bunch of fishermen at once, you won't be able to tell their religion or their politics. Which is maybe a good thing. But they reveal their interests when they talk, and of course fight among themselves, about different gear types and fisheries and so on.

I have tried to figure out the biggest differences between fishermen and the biggest one I've come up with is smoking. Some do, some don't. So I got to thinking about that and about health. How many fishermen are healthy? It beats me.

All this is on my mind because I have been hanging out with my hiking and fitness friends. Some of them are vegetarians and most of them are against fried foods and smoking. As my body gets older I want to tune it up, and after lying around all winter I want to get fit. Too much bunk time does not a good fisherman make. As a woman on a boat I am especially under the gun, and there is always some comment about my strength and stamina. So if I want to get a job on the salmon and keep it, I have to be fit. At least that's my theory.

I've been walking along the dyke into Steveston and back just to start the blood flowing, and lately I've been keeping an ear to the hikers to get all the latest. At first, all I got was the obvious: walk lots, no fat, eat your veggies and so on. I also got all the weird remedies such as the ginseng tea, the vitamin diet and everyone's favourite recipe using lots of garlic. I listened to all this and kept asking them who was the fittest person they knew and what was their secret. I heard a lot of names but one name kept coming up.

The guy's name was Mike and he made his living as a survival guide. He took tourists into the wilderness, showed them how to survive there, and charged top dollar for doing so. Obviously, to do his job he had to pack lots of gear and be very fit. Evidently he was not a young guy either. He had been in the army as a commando and took early retirement so he must have been about fifty.

Nobody knew too much about his personal life, but they all knew about his abilities. For example, he had his own survival pack containing, I guess, a gun and a first-aid kit, as well as survival gear, emergency food, and his sleeping bag. This, I was told, weighed twenty pounds. So where was the big deal? Well, he strapped that to his back, along with his client's pack which was about forty pounds. So on every trip he packed sixty pounds into the wilderness. This is how he made his living. I heard so much about the guy that I had to meet him.

What a shock. He was no Greek god. On the contrary he was rather ordinary looking, needed a haircut and a shave

Two Matches and an Attitude

and had a beer gut. That was bad enough. Then he pulled out a pack of cigarettes and lit up.

This was the fittest guy around? Well, evidently he just happened to be one of those rare people who are naturally strong and have all the stamina in the world. I was so disheartened I didn't tell him about my efforts to get fit. I was actually there with some friends of mine who wanted to hire him for a wilderness trip and needed a list of stuff to take. They were going to put their tent and food in a pack and get Mike the guide to carry it. They wanted to minimize their weight, so a list from him was necessary.

After the business was done I got to shooting the breeze with Mike. He told me some war stories and then some horror stories of clients he had. He was a good storyteller and I could see why he would be fun in the wilderness around a campfire. He was an authority on survival techniques and told me many interesting things and gave me some tips if I was ever lost or shipwrecked. But the most important thing he told me was his motto: two matches and an attitude. Clothes optional, he added. The matches were necessary for a quick fire to ward off hypothermia or get hot food. The attitude was the key to it all. With it you can survive in the wilderness, without it you will die. I had to believe him because he made his strange living from this philosophy, even though he shattered much of the conventional wisdom about fitness.

Thinking it over later, I realized that what the guide had told me was also important for survival in the fishing industry. I didn't need the two matches but the importance of attitude got my attention.

The fishing industry is going through a fundamental change. The question we all face is: will I survive? Too many boats, too many people, not enough fish. Big changes in licensing and fishing times are the order of the day. My hold on the industry has been getting weaker each year. From the joint owner of a small seiner a few years ago to just a personal licence and a gillnet herring roe licence means that the tide is against me. I don't want to end up as another statistic. So attitude must see me through.

The other thing I learned from the guide was that the usual rules don't hold. He smoked, he ate fried foods, he had a beer belly and yet he could outwalk you with a sixty-pound pack on his back. That made me think: so what if I'm not strong, I don't have a boat, I'm a woman? Boy, do I have an attitude. So to heck with the ginseng tea and that other stuff from my hiking freaks. Gimme the fried food and a couple of beers. Lots of sockeye coming this season and I've got the attitude to catch them.

Bull and the Boys Come to Town

Finally—a couple of leads on jobs on the salmon. How I got them is quite interesting, as they just kind of happened by themselves.

I was doing some bookkeeping work for some friends who are rebuilding seines in the Paramount net loft. They were paying their crew wages and needed the exact payroll deductions for the government stuff, especially Unemployment Insurance. The salmon season is so short that everyone needs all the stamps they can get. To make sure my calculations for the paycheques were correct, I took it over to my friend Jenny. She is an accountant and doesn't let you call her a bookkeeper.

I went over to her house, which is quite nice. She works and so does her husband, as a schoolteacher. He used to be a fisherman and that is how I met Jenny.

When she answered the door I saw trouble. "Bull and the boys from the Bay are here," she hissed. I smiled at the alliteration, but at the same time there was cold fear in my stomach. Bull was here. The "boys" were no problem—they were merely his brothers and cousins and whatever—but Bull was a problem. Surely someone somewhere knew the name his mother gave him but everyone else knew him only as Bull, which was short for "bull of the woods," which is a BC term for someone who is what the anthropologists call the Alpha Male.

Anyway, I went in, but Bull was not there. He was coming to see Jenny's husband for some reason or other and had sent his pals over first. Jenny was steamed because there was no prior phone call—they all just showed up with several cases of beer and were waiting for Bull.

Jenny and I went into the kitchen to talk while her husband and the boys sat in the living room. The "boys" were all grown men, and all appeared to weigh about 300 pounds and looked like seine skippers, which they were. To know one was to know them all because they were all related to each other. While we talked we heard a car door slam. I peeked because I knew it was Bull. Leaning against the taxi was this man well over six feet, who had dined often and well and could never again get into Extra Large. He had the physique of the very strong, a massive chest and long, sloping shoulders. Atop this frame was a face that was too small for his body and was strangely boyish and innocent looking. He was struggling with three items in his arms and hands and since, despite his size, he had the usual number of these, he had a problem. Getting himself organized, for he had been drinking, he threw himself at the house.

Once inside, the three items Bull struggled with turned out to be a case of beer, a bag with empties and mixer, and another bag with a half-empty bottle of cheap rye whisky in it. He offered rye and ginger to all and all declined. He handed the beer to Jenny without a glance and said for her to put it in the fridge. Then he opened a beer and they all settled in to talk. Jenny and I remained in the kitchen and finished our business. But Jenny was steamed and I came to the horrible conclusion that she wanted to have a go at Bull. It was not long in happening—Bull gave her an excuse.

"Tell your old lady to get me a beer," he said to Jenny's husband.

She stormed into the living room and lashed out. "I'm not his old lady and why the hell can't you guys phone ahead when you visit?" she demanded. "Get your own beer."

Silence all around as Bull looked at her. Jenny's husband cringed. The boys watched. Very quietly Bull said it was women's work to get their husbands' friends a beer. Jenny replied that she wasn't on Bull's boat and didn't have to do as he said. Bull replied very emphatically, "Women are no goddamn good on a boat."

That tore it for Jenny. "Your brother likes women," she said. "His wife cooks on his boat, and your cousin has two of his daughters on his." She fixed Bull with a glare. "And your mom cooked on your dad's boat, was she no goddamn good?" Silence. Deep, deep silence. Screwing up his face to signify that he was thinking, Bull gave his summary.

"Facts are facts," he said.

This, of course, was no answer at all, but it did the job. Conversation resumed, but Bull had been challenged and Jenny appeared to have won. But Bull was biding his time. When they were all leaving and getting into the taxi, Bull suddenly addressed Jenny for the first time since the flareup.

"Hey," he said, "why do married men die before their wives?"

Jenny fell for the bait and said she didn't know.

Bull replied. "Because they want to," he said. "Because they want to."

Laughing uproariously, he fell into the taxi knowing he'd won. And all Jenny has for her trouble is the remains of a bottle of rye and some beer bottle empties. Bull and the boys went home to get ready for the salmon. Unknown to Bull and Jenny, in the meantime I talked to the boys and got a good lead on a job on a troller, and Bull's brother's wife has to go into the hospital—probably because of her husband—and I may get a job cooking or running the drum on his seiner. I know I won't get a job on Bull's boat, but one way or another, salmon season is starting to shape up.

Setting Your Net

I was over at the Annieville Slough and who should I meet but Barry, the red-headed deckhand who used to work on our old seiner. It seemed so long ago, but I was nervous around Barry because you never know what he'll say or do. Also, he used to drive me nuts and I didn't want old sores to be opened up. I thought he would start asking me about my marriage breakup or something else that was none of his business.

But Barry had mellowed. He was married now and he owned a troller, and soon we were exchanging information the way fishermen do. If he gave me a piece of information that was new and useful I would give him something. I had to admit he was way ahead of me, so we turned to more general subjects. He starting talking about the problem of making a troller a fishing machine.

Every boat is different and the essence of trolling is to make the boat fish to its maximum. The length of leaders, shape of flashers, swivels, plugs, and the whole grab bag of stuff that trollers have would boggle the mind if you tried to explain it. And there is always that nagging fear that you haven't got it right. In trolling, there are always a million adjustments you can make.

Gillnetting for salmon is different. Everything depends upon how you set your net. It takes some knowledge and lots of guts, but the more knowledge you have, the less guts you need. Johnstone Strait gave me the willies but for some I am sure it's like a walk in the park. The technical problem there is that you drift with the tide and it's all done at night. Where you set your net is important. Keeping it straight and fishable requires constant attention.

For added excitement you have log barges coming through doing up to 20 knots. Once in a while you get it perfect, a slight westerly giving the sea a small chop and a bright moonlit night to get the fish up. No "fire" in the water and lots of fish gilled.

Setting a salmon seine is a different proposition. Ideally, you want to tie up to the shore (always called the "beach") at a traditional setting spot, and you'll get a nice tide and of course lots of fish. Timing is the key. You want to have your net out when the fish are really coming. Nowadays there are long lineups at the good spots, so you set when your turn comes and hope for the best. The trick is to get into some sort of routine or rhythm so that you are setting your net in different places at more or less the best time for each place. When it happens it is pretty.

The alternative is not pretty. A sudden tide surge or a backing up tide can cause problems and you snag up. Usually you can drum off a net snagged on the bottom and get away with only a few rips and tears. That is, you use the hydraulic power of the net drum to pull the net off the hangup. Other times you may leave your lead line behind. No net, no fish, no money.

There are a number of other treats in setting a seine. Getting it off the drum can cause rips and tears if you get the dreaded backlash. Then there is the wrist-strengthening drill called "getting the roll-up out." This happens when the net rolls up in a tight ball and has to be unravelled. No high-tech stuff here. Just muscle and elbow grease for fathom after fathom. The welcome words are "Here's the key," for every roll-up has a key and once found the roll-up can be unwound and the net is fishable again.

Once the net is set, how you tow it depends upon tide and power and where you are fishing. Fishing for pinks in the northern inlets is easy, as they are slow moving. Sockeye are something else. You do short tows, make a hook in the tow end of your net and close up without towing so hard you lift your lead line. This keeps the sockeye in the net. So much to know, so little margin for error, and no manual to consult when things go wrong.

I know all this stuff, as I keep records of every set we made while gillnetting and seining. It is never enough, as the time and tide and date and spot are always different, but the knowledge helps. There is a payoff for experience.

So what does all this mean? I really believe there is a connection between how you set your net and what you need to know in life to get by. There is always danger, things will always change, you can never predict what the results of your efforts will be, and all you have is your experience to draw upon.

If they can learn all that, it seems to me that fishermen should surely end up as the most successful of people, or at least the richest. Just why they aren't I don't know, but I do know that the challenge for me is to take the skill necessary in setting your net and somehow be just as skillful in life. Set your net right and you get fish; live your life right and you get, well, something. It's tough to figure out just what that something is. That's why in spite of all the difficulties, fishing is easy. At least you know what you want.

Adams Fish

In fishing, every year is different, but as in other aspects of life there are a few cycles you can depend on. Every four years we salmon fishermen get something special: the Adams River sockeye run. And every time the cycle comes around we ask the same question: will it be like 1958?

I didn't fish in 1958, but every time you discuss the Adams run, someone within earshot will remember that year. The run was big and everyone had a crack at them. The gillnetters got them, although the Adams sockeye are hard to gill when they are in big schools. The seiners loaded up, both in Johnstone Strait and off the west coast of Vancouver Island. It was the year that finally taught the trollers how to take sockeye on their hook and line gear. The key setup was "a hootchie-kootchie and a kelp-cutter." Trollers argue about the best colour for the hootchie-kootchie but non-trollers are lost. We have little or no idea what the hell they are talking about.

Last time the Adams run was due, we worried about it.

How big a run would it be? As always, under some formula or other the DFO had allocated a percentage of the catch to all the gear types, so that everyone would get a crack at the sockeye, but that year promised to be somewhat different. Our so-called "fish war" with the Americans had caused a fair amount of turmoil. As I understood it, the fishermen in Washington state and Oregon want more of the sockeye and we might give it to them, but in return we Canadians wanted the Alaskans to take less of our fish. The Americans couldn't agree among themselves. Canada wanted the Americans to get their act together as the Adams fish were really our fish. The negotiators called it "the interception problem." We fishermen called it something else.

In the midst of all this confusion I was hoping to be filling my pockets. The Canadian fish war strategy appeared to be to fish intensively off the west coast of Vancouver Island, where most of the sockeye would likely pass. That meant a job for me if I played my cards right. If I didn't get a job on a seiner I would be dickering with Uncle Ed to be deckhand on his troller. At least he knew we could work together, an important advantage as Ed gets older and crankier and has more trouble with his crew. When he was young he only needed one deckhand, but now he needs two as his knees are giving out on him.

I knew we would be getting a poor price again because of the big Alaska fishery and the competition from the fish farms (strange, not only do we have too many fishermen, we have too many fish!). Costs still seemed to be going up. Boat prices are always a mystery to me but they seemed to be falling. Maybe after that season I would have enough bucks for a down payment on a troller or gillnetter. As a woman in the industry, the potential for a permanent job on a seiner was getting remote. Working on deck running the drum was the job classification I aimed for, but there were only 549 seine licences, and 3,000 men who could also do the job, so the odds were not good for me. Those were the hard realities I was facing. But Adams River sockeye allow you to dream.

The beauty of Adams River sockeye is twofold. They are beautiful to look at and they are beautiful to fish. They travel in big schools, and if you are skillful and lucky enough you can have a big season. They are also fun to fish, as they are wild and tricky and not everyone can catch them. To me they sort of capture the magic of fishing, the excitement that comes from mystery and tension. There is a big mystery and a small one. The big mystery of the salmon is why and how they come home to die in the very stream they came from. The small mystery is being out there to catch them, and never knowing how or when they will show up. The tension comes from never knowing if one sockeye jumper in your net means a hundred more fish or ten thousand. Sockeye fishing is an adrenalin pumping exercise.

That year, with the complication of the fish war, and the prospect of more fish wars to come, we were all confused and worried.

But the salmon were not worried. They have a simple mission: get home. This means that every four years, in early August they get close to the Fraser River and the Adams tributary. They go through Canadian and American waters, avoiding nets and lures, pursued by seals, sea lions and killer whales. Once in the Fraser they must go through Hell's Gate at just the right time to enter the Adams River. Every four years they come and whenever they come they make it. I only hope that I can too.

Homemade Philosophers

Worry, worry, worry. Summer is here, and as the Japanese lady poet Sei Shōnagon said a long time ago, it is the nights that are the most beautiful. Last summer my nights were troubled because I knew the salmon were on their way, but where was the fishing time? No fishing time means no fish and no fish means no money.

At least I had a job. When I helped Frankie the Nontalking Deckhand buy his own boat, little did I know that I was sealing my own fate, for what do you know, then Ed was in need of a deckhand. For women in the fishing industry—as for everyone else—the best way to get a job is from someone in your family, so I became deckhand on Uncle Ed's troller. There would be few sockeye and fewer springs and we weren't going to make a dime on pinks ("humpies"). But we needed time to catch them. Worry, worry, worry.

I have always been a bit of a worrier and it shows. Ed, on the other hand, is just plain cranky but not a worrier. Well, I know he worries, but he kind of does it all beforehand

Homemade Philosophers

and then makes a decision about whatever it is and then won't worry any more. For example, Ed was one of the first to build a big troller, and everyone said it wouldn't catch fish and so on. Ed worried about it all winter. But once the boat was in the water he didn't worry any more, and told everyone it was just the ticket for him. I know that first season wasn't so hot for him, but he never blamed it on the boat. Once he made the decision, no more worrying. He was right in the end, as around him he saw trollers being built bigger and bigger.

Thinking it all over—I mean my worrying—I concluded that fishermen have to be philosophers of some sort or other or else they will go crazy. Like most people who have been to university for a spell, I got my philosophy out of books. None of it made much sense to me so I bought *Philosophy Made Simple*. Still not much I found useful, but I do recall packing around philosophy books hoping people would notice. I never read them, of course. Philosophy out of books and from a university never helped a bit in the fishing game. As a matter of fact, I never thought about that kind of philosophy again. Fishermen do need a philosophy, but they have to make their own.

When I go to Steveston to the boat or to the net loft to get something out of the locker, I make a payload of the trip by phoning some of my pals to meet them and exchange news and gossip. And I know that along the way I'll get some kind of philosophy. I call it homemade philosophy and I guess my pals are homemade philosophers. Homemade philosophers got that way by being fishermen and having to deal with the assaults of weather, cold-hearted bank managers and an industry that is having its problems.

We are quite a crew when we gather in the Net Shed Cafe. We often see and hear the gillnetters shooting the breeze, but they only seem to talk about golf and whoever is absent. No philosophy from them. We trollers are different. We have the same sort of problems, but we look for the hidden meaning or at least try to get a fix on the problem. I guess trolling does that to you. Too many long tacks with no fish get the mind going.

Any event or circumstance is bound to force the brain cells to fire up and agitate the emotions. We go from the concrete event to the abstract concept. We got fish here yesterday but there is none today. What is the meaning of it all? Is it fate, karma, kismet, or just plain bad luck? Or does it have nothing to do with that but is merely that the water temperature has changed? Real philosophers go from the abstract idea and invent a philosophy. We do it the other way around. We try to get a philosophy that will suit us.

Being a homemade philosopher has its rewards. First of all, you develop an attitude that allows you to handle any sort of trouble. From a cold cup of coffee, right down to trouble so horrible that nobody ever mentions it except in the small print in insurance policies. Fire, earthquake (Steveston turns to soup), shipwreck, acts of God and even the restraint of foreign princes. Not even to mention $2.50 a pound for large red springs. Or the possibility of the shortest fishing season in living memory. One day in that worrisome season, I bugged each of my pals to tell me his philosophy. Here are the results:

Uncle Ed: Do the job of the day on that day.

The Scientific Fisherman: Use the scientific method even though it's not scientific to do so in fishing.

Barry: To spawn you have to swim against the current (said with a grin).

The Non-talking Deckhand: Talk is cheap.

A guy in the net loft at Paramount: Always have your gear in the water at low-water slack.

A seine skipper: A good skipper makes a good crew, except when he doesn't.

A worker at Trites shipyard: Pay your bills and you can tell everyone where to go.

Now, none of this is really philosophy but it will do in a pinch. It seems to me that fishermen take whatever is at hand and apply it to the situation and if it works, sort of, then it's okay. So it is with a homemade philosophy. Here is something that works, so it is my philosophy—at least until new problems come along and I need to get a new one. Of course, if we had lots of fish and good prices then we wouldn't have to make up a philosophy. Homemade philosophy is the only good thing you get out of no fish and no money. I'll take more fish any time.

Jonah in the Belly of the Whale

I don't know how many Bibles there are on commercial fishboats but I'll bet most fishermen have heard the story about Jonah and the whale. First of all, the word or name "Jonah" is important. Jonah was thrown overboard and swallowed by a whale because the crew on the boat thought he was unlucky. To this day all bad-luck fishermen will say they are not superstitious but they are to a very high degree. And a Jonah is one of the things they are most superstitious about. Sad to say, but some people are considered bad luck and are whispered about as though they have leprosy. They are the unfortunates who are Jonahs.

Uncle Ed has stretched the concept a little further and considers some boats to be Jonahs.

The seasons are getting shorter and shorter and starting later and later. And worst of all, the fish are now getting more scarce. The chinooks—what we call springs or smileys—are declining, and the last time we got out fishing the Alaskans were being naughty. They said they had conserved

Jonah in the Belly of the Whale

long enough and so they shouldn't have to cut back on their interception of our fish. Impeccable logic if you are an Alaskan, but a bitter pill for us to swallow. But Uncle Ed blamed the tough season on a boat that had been following us on the grounds—a boat which, to Ed's mind, is the worst sort of Jonah. A boat that gives you bad luck for a whole season.

The problem with bad luck is that you have to get rid of it. Now, some Italian friends of mine have given me some good cures against the Evil Eye but they were not effective here, nor were certain other highly secret measures I have at my disposal. No fishing time, a lousy price, and a Jonah following us around caused me to reflect on the story of Jonah and the whale.

The story itself is highly improbable, as you may recall. Jonah was thrown overboard because he caused a storm. He was swallowed by a whale. While in the belly of the whale he prayed, and as a result he was disgorged. Out of the whale, he once again was safe and sound.

My goofy idea was that if I could figure out what Jonah was thinking about while in the belly of the whale I might come up with something that would get us out of our bad fix. Surely, this idea of mine was no crazier than the thoughts that Jonah must have had while he was in the whale. He must have explored many ideas before he was reduced to prayer and faith alone to get out.

The more I thought about all this stuff the more I became obsessed with it. I knew I couldn't just ask Uncle Ed what did he think Jonah was thinking about while he was in the whale. Ed would just snort and swear at such a stupid question. To take advantage of Ed's thinking, I had to ease him into the discussion.

The first step was to discuss why Jonah was thrown overboard. Ed saw that as an easy one. Fishermen believe in cause and effect. Jonah was new on board the ship. Bad weather came, therefore he must have caused it. His companions' solution was the right one: throw him overboard.

The next day I got Ed to talk about the whale by talking about whales in general. What was the biggest one, the most in one spot, the closest he ever got to one, and all that. Then why the whale swallowed Jonah. My technical arguments about a large whale eating only plankton and not able to swallow Jonah got nowhere with Ed. Jonahs must be punished! Swallow him the whale did. Now came the touchy part where I wanted Ed to do some thinking. What was Jonah thinking about before he resorted to prayer? Ed's first answer was the obvious one: how to get out. Now I had my opening. I belaboured poor Ed on the plight of the fishing industry. How do we get out of the present fix?

For some reason or other this struck a responsive chord in Ed and strange to say we had a couple of good days discussing the industry—how we got into our situation and how to get out. I like Ed when he is in a mood to come up with insights and observations that you would not expect. Also, Ed can argue for policies that are not in his own self-interest, which is surprising for a fisherman.

We got so caught up in all these thoughts and philosophies that we almost forgot about the Jonah boat that was following us. I guess that did it. We weren't thinking about him and bang! he was gone. Mentally, we had thrown him overboard.

A couple of days later, Ed asked me what that business of Jonah and the whale was all about. I explained the industry right now was like Jonah in the belly of the whale. We had done something wrong, had been cast overboard and were swallowed by the whale. While in the belly we had to really do some thinking on how to get out.

Ed gave me a look he usually reserves for politicians and idiots. Sort of like a peculiar odour had entered the room. I knew Ed disliked arguments that he calls highfalutin', whatever that means. But I knew Ed would think it over. He doesn't like being in the belly of the whale. Neither do I. Whatever else there is to be said about lousy fishing, it can sure excite the brain.

Tom the Skiffman Gets Dressed

The Blue Line is what we call the boundary limit between Bonilla Point on Vancouver Island and Tatoosh Island on the US side. You're not allowed to fish beyond the Blue Line, named for the blue line in hockey which you can't cross without the puck. Because of this, it gives the seine fleet their first chance to catch Fraser-bound sockeye and pink salmon as they head back home from the high seas.

Fishing the Blue Line as I did recently, I had enough adventures to satisfy me for the next five years. It was rough and tough. After all, once you get to the Blue Line you are off the west coast of Vancouver Island, right on the open sea, next stop Japan. We had bad weather and bad tides and all the other stuff to test us, plus no fish to speak of. But I noticed lots of things that surprised me. There are certainly more women fishing on seine boats than I expected. Also, more small boats out there than one would think. But for my notebooks, I opened a new section on what fishermen wear

Tom the Skiffman Gets Dressed

when they are working. One of the crew, Tom (not his real name) the skiffman, was my subject.

Tom is your average skiffman, about five feet ten, and says his fighting weight is one-eighty. He appears to lead a full life: his pals in Steveston call him Roadkill. What that means I didn't ask. Because the weather was so variable—fog, wind, rain, sunshine and you name it—I paid great attention to what Tom wore as the day progressed. After two weeks out there, I think I know what the well-dressed skiffman should wear.

Let me explain a bit about what fishermen wear. On shore, they wear just about anything, and at sea sometimes that is also the case. But skiffmen are in and out of the skiff, have to be able to keep warm and dry, or cool, and also have to be able to move about with some amount of speed and skill. The alternative can be nasty. Falling overboard after being hit with a snapping line in the Strait of Juan de Fuca is not a prospect even I would enjoy. So Tom would dress sensibly, with care, and prepare for all eventualities.

Tom paid attention to two articles above all else: his Polaroid sunglasses and his life vest. The reason was obvious—the glare of the sun and drowning were to be avoided. I assume the golf club hat was a matter of personal choice. Much more interesting was Tom's selection of footwear. Dry feet are the key to a fisherman's happiness. Tom had some special high-tech socks that promised to keep his feet warm even if they were wet. Not that his feet ever got wet. His rubber boots had been selected for their ability to grip the deck (aluminum) and for climbing in and out of the skiff.

Choice of rain gear was also important. Tom was a traditionalist and stuck with Helly-Hansen pants and parka, but some of the crew used gear that climbers and skiers favour. Under his rubber gear, Tom applied the layer principle. As the day got warmer, he shed an article of clothing and then put it back on as night approached. Now, I never actually saw Tom get dressed, as I wasn't in the big stateroom. For whatever reasons, I bunked in the cook's room, he in the top for the porthole and ventilation, and me in the bottom from lack of seniority. But on a seine boat, there are few secrets, and I knew Tom's outfits as well as my own. What Tom wore and what I wore were highly similar, but he took more care. We did, however, wear the same type of gloves. They are yellow-coloured and are a mixture of cotton and plastic that make them ideal for fishboat work.

There are many different kinds of work on a fishboat, and what fishermen end up wearing is the result of practical choice. As the work changes, the gear changes, and over time certain things become essential. The engineer, "Chief," wore the black Romeo slippers treated to resist oil that have been in the industry for a coon's age. He also favoured the wool pants and suspenders used by many an old-timer. Not one fisherman in ten knows that they are called Bannockburns, but so what.

I guess Tom the skiffman was in the middle between what the Chief wore and the jumpsuits and high-tech stuff the more up-to-date types put on. No one talked much about what they wore or why. Fishermen are practical people—doers rather than thinkers. No one gave a damn when I told them about the checkered design on their wool work shirts (with the sleeves cut off to avoid saltwater boils and buttons hanging up in the net). The design has evolved from traditional shirts worn by Basque fishermen, and it has somehow migrated to Canada, probably through Newfoundland.

I find this stuff interesting and I keep lots of notes. But some things I don't know what to do with. When I complimented Tom on his care and skill in selecting clothes, he merely snorted, "Hey, a man has to get dressed, hasn't he?" And so he does. I also said to Tom that I thought he always looked healthy. After a pause, he noted that he did have an attack of arthritis in his shoulder last year. I asked how he cured it, and almost fell overboard at his answer. He rubs it with something out of a spray can called WD-40, which the Chief uses to free up rusty bolts and lubricate everything that might seize up in the engine room! In this fishing racket you have to be like Tom—ready for anything.

Fishing in Roll-Up City

The Pacific coast drum seiner is the slickest way in the world to get fish aboard a boat—when it works. One year my husband and I were fishing up around Dundas Island, chasing Nass and Skeena fish on our little seiner the *Fantasy Isle*, and it was tough.

Things worked well on the first day, but on the second day disaster struck in the form of roll-ups.

Setting a net from a drum seiner requires a bit of skill on the part of both the skipper and the drum man who controls the brake on the drum. When you are ready to set, everybody stands by and the skipper hollers "Let her go," and the skiffman drops the sea anchor in the water, and the end of the net and the skiff are cast adrift. The net then starts to come off the drum. The drum man stands by the brake, ready to stop the drum if two things happen: a backlash, or the lead line jumps over the cork line.

We have not had a backlash, but lots of lead line jumping over the cork line, caused by the skipper setting the net in too tight an arc. By applying the brake, the cork line tightens up and is supposed to flip the lead line back over on the right side.

On our first day, we knew all this and set the net with no problem. And even got fish. When it all works, it's a joy to fish. So on our second day we flew at it, but on the third set I noticed something funny about our cork line when we were towing. I thought some seaweed or kelp was stuck on it.

When we drummed in, I saw that some of the web had rolled around the cork line. "Uh oh," said Barry, who was our skiffman in those days, "here comes roll-up city." I was a rank beginner at drum seining and I didn't know what he meant, but I do now.

Roll-ups are the curse of seining and especially drum seining. That small cork line roll-up got worse each time we set. More and more web kept getting wound around the cork line, until it was a great big black ball about four fathoms long. Your net does not fish at that point.

By early afternoon, we had had it. We tried to fix it on the grounds, but it was no use. We had to go in early. It meant valuable fishing time that we couldn't afford to lose ... anyway, it was into Rupert and onto the dock, to twist the hell out of the net with our bare hands.

You cannot believe how tough a roll-up is until you have to fix it.

You have to wrestle several fathoms of cork line with several strips of web tightly rolled around it, and you have to twist and get slack until you find the ever-elusive "key" that will let you start unravelling the whole mess. It took us hours.

That was our first week. The second week started off just as bad. By ten in the morning we had another cork line roll-up. The choice you face is cruel and simple: stop and fight the roll-up now and lose fish, or leave it alone and hope it doesn't get worse. But it always does.

We left it alone. The tide was flooding and we were getting fish.

Every time the part of the net with the roll-up came off the drum, Barry would holler, "Roll-Up City, here we come." If I'd had a gun I would have shot him. And I told him so.

He replied that he had never seen a woman who was a good shot, so I would probably miss. I gave him a look, and he just gave me his foolish grin. But the roll-up never got worse. It only took us an hour to get it out on the weekend.

Rupert on the weekend was and is a crazy place. Every seine boat on the coast must have been in Rupert that weekend. They were buying up all the grub, filling up the beer parlours and bars, and carrying on as only fishermen can. If you like BS, gossip and character assassination, then plan a Rupert weekend as part of your summer vacation.

And there was plenty to keep us talking. A strike coming up, the trollers wanting to fish sockeye in Johnstone Strait, the gillnetters trying to get a special new area, Alaska's Bristol Bay fish were or weren't coming in, and on and on it would go.

After a month, I wondered that we were still alive, although in reality we had done better than we had hoped. But the frustration and worry of this kind of fishing were more than I could have believed. The continual strain of always fighting other boats for position and for fishing space. The tension among the crew, as we were all strangers to one another—and Barry in particular always got on my nerves. It's taken me a good ten or fifteen years to get used to his funny little ways.

But there were some sunshiny days. And one day, the

roll-ups just went away. Where they come from and where they go is one of the mysteries of the universe.

Fish and Berries

Every spring the question is: are there going to be lots of fish? and every summer we find out. My life has always been that way. When I was a kid we used to ask the Indians. In some areas of knowledge their authority was unchallenged, and a question like that was right up their alley.

The one indicator that everyone looked for was the amount of salmonberries. If there were lots, then it followed that there would be lots of fish. There's nothing more exhilarating to a fisherman than lots of fish.

For example, the craziness called Johnstone Strait during one of those magical, once-every-four-years Adams River sockeye runs. I had been in the Strait gillnetting before but seining is different, and those fish are something else. They move fast and deep, travelling in big schools. Catching them takes skill, brains and luck, and even if you have all three of those things, you still may not get them.

When there was an opening we all anchored up for

Fish and Berries

beach sets, or cruised offshore for a spot to set on a good school. When six o'clock came, all hell broke loose as every net hit the water. A couple of hundred seines in the Strait is a sight to see. And the radio! You wouldn't believe what grown men can say and how they will act. But the stakes are too high for the Marquis of Queensberry rules. It's every man for himself.

After the first set it quieted down, as everyone examined the results of their sets and reassessed their overall situations.

We usually leave the Strait after the opening night as most of the fish are mopped up. Some skippers like tide sets, where you wait for the tide and then tie up your net to a rock or tree on the beach, but we are usually too impatient for that, so we head for the north shore of Malcolm Island.

The fishing there is all offshore, and we set and set and set again. It has a pattern. The fish come on the tide, and sets are a little better than before, then they drop off after high water. Late in the afternoon a westerly comes up, and you have to drum the net in a certain way so that you "make the corner." This means you drum so that the wind blows you out of your net just as you come to the last forty fathoms. If the wind blows you into your net, then you have a hell of a mess of gilled fish and snarls. Lost fishing time is a sin here in the "sockeye races."

Fishing these sockeye is a frustrating and tense proposition for seiners. Usually the problem is not enough fish or too many boats. During the Adams River sockeye run, there are lots of fish—but just try and catch them! Those who are used to fishing lazy humpbacks (pink salmon) in northern inlets get the shock of their lives. Lots of fish go into their nets, but they hustle right out again.

The reason is that these fish are in a rush to get to the Fraser, and they aren't going to wait for some dumb fisherman.

You must set right on these fish, and soon as you think they are inside, you must close up. We put a hook in our towing end immediately, and if we see a jumper or finner inside we close up. This means you cut off other fish coming into your net, but you have to do it. If you are a "log-tow artist," and stay open as long as fish are coming, you will only get the last two schools.

We learned this the first week that I seined an Adams River run with my husband on the *Fantasy Isle*. We weren't getting many and our deckhand Barry, in his polite way, said that we'd "starve to death trying to catch northern humpback in sockeye country." After a bit of shouting between Barry and my husband, we tried Barry's advice and it worked.

Fishing to beat hell, getting them sometimes and missing them other times. Every set was a gamble, and I trembled every time we let go. But nothing beat the thrill of "making the corner" when the westerly blew off Black Buff, and seeing sockeye bubbles in the net. We got them again—oh no, we missed—well, try again. That's the fishing life.

The next weekend, we tied up in Port Hardy and went berrypicking. Salal berries were just starting, and some Indians had told us there were going to be lots of blackberries. It turned out they were right. Lots of berries, lots of fish—what a life!

Tide and Currents

Back in Steveston between trips, I ran into my pal the Scientific Fisherman on the gillnet floats. I surprised myself by being glad to see him. Scientific in name but not in practice; I have hardly ever seen him apply any of his so-called knowledge, so I tend to get tired of listening to him after a while.

However, this time I caught him out. He was on the float with a couple of new charts and the tide book. With a couple of paint cans and a gillnet float enacting the roles of the Earth, the Moon and the Sun, Scientific was showing the difference between a big tide and a small tide. Pretty basic stuff, but his audience was enthralled.

Because most fish are dominated by the tide, so are fishermen's lives. For example, which way the current is going at any given time is important, although we never say "current"; it is always "the tide is right" or "it's good." Sometimes we will be a little more specific and use the words "ebb" or "flood." More likely we will say "rising tide" or "falling tide."

141

When discussing a fishing venture, where to go, when to go, what to do, everyone makes a mental image of the state of the tide, as this underlies everything that will happen on the trip. It becomes second nature. For example, leaving Steveston for Prince Rupert you make a mental calculation of the time it will take your boat to get to Seymour Narrows. You leave on the flood, get to the narrows at low water slack and enter Johnstone Strait at the start of the ebb. Why? Well, that way you will have a fair tide all the way and make good time. Also, being in Seymour Narrows is an exercise not easily forgotten.

One skipper said to me: "Going through the Narrows for the first time when you are in command of a boat is like holding a wildcat upside down, with one hand, by its tail. You gain experience and wisdom that you can get in no other way."

In other words, you need to know your tides and the currents they cause. You can get to know them in detail by years of experience and memory, or you can buy the book and look up its tables. The BC coast is organized into areas with reference points for the times of its tides. You look in the book, see when high tide is at Such-and-Such Rock, and do some simple mental calculations. You need to know if the current date is one of big tides or small ones. A full moon brings with it a big tide, which is when there is a large difference between the high and low tides.

Every fisherman knows that there are profound differences in the way tides act at different places along the coast. In Vancouver the tides are relatively small, whereas in Prince Rupert the differences are huge. Also, fishermen know that some times of the year tides are big and other times they are not. It can be confusing, but it is all in the tide book for you to figure out. One important point—*do not try to read the explanation of what causes tides and the various differences between them.* It is written by the guy who writes the operating instructions for Japanese electronic equipment.

That is why the Scientific Fisherman was using paint cans to represent the various heavenly bodies. Fishermen know the sun and moon cause the tides but they are not exactly sure how. One big puzzle is that in Johnstone Strait there are two definite tides per day. However, if you fish the west coast of Vancouver Island there only appears to be one definite tide a day and then a strange kind of mini-tide—sort of one and a half tides rather than two.

If trying to figure out tides isn't bad enough, there is the problem of figuring out currents. The tide moves big chunks of water around and this causes currents. The timing of these currents is important. Going through a dangerous place such as the Seymour Narrows or any other narrows is one thing, but the current has an effect on fish, especially salmon. Salmon are not dumb. When the tide is against them, they lie low. When the tide is good, they move. This is when fishermen are able to catch them. The time that the current changes depends on where you are in terms of location and where you are, timewise, in the lunar cycle, which determines the size of the tide. Good fishermen write these things down, because each year the tides are more or less the same, so the currents are too, and so are the times that they change.

I guess I have read the tide book a thousand times and yet I really cannot give a good explanation of how the tides on the coast act and why. I know many small things, such as when a particular backeddy will form in a bay because of a large tide, or how many minutes after high water the current will change, or which of the many things you should do to catch fish in relation to the prevailing tide. But the grand picture eludes me. I got the books out from the library and I am no wiser, except I know that the tides are exactly the same every 18.6 years. Moon and tide time will be at the same calendar time every 93 years.

The lesson here for me was in what Uncle Ed said when I laid all this out for him. He is of the opinion that the fishing industry is in a period of great change, but that every 93 years it will be the same. His logic escapes me, but I get his point. Sometimes the tide is against you, sometimes with you, but if you wait long enough a time will come when you'll know exactly what the right thing is to do. You just have to be prepared to wait for 93 years.

Turn Your Radio On

I thought getting back to trolling would be fun because it was what I started out doing in the fishing racket so many years ago with my dad. Well, it wasn't that much fun, and not because there was no price for our fish, but because the feeling wasn't right on the boat. There was that boat that followed us on one hand, and the many stimulating exchanges that Uncle Ed and I shared on the other. The boat is a highliner and has two crew. I had to do the cooking, which I don't mind, but because I am a woman I still feel, even after all these years, that Uncle Ed and the other crewman just plain didn't respect me and my skills. If I lost a fish they just looked at each other. It made me mad because I earned my keep and taught Ed a thing or two whether he'll admit it or not.

But it can go the other way. I went on a seine boat once where they shuffled the crew a bit and I wasn't the cook. I ran the drum because on the west coast of Vancouver Island the engineer, who would normally do it, is busy with cooling fish and when bringing in the net just stays at the hydraulic controls for the boom, pursing winch, Gearmatics and other stuff. But we didn't fish the west coast on this trip. It didn't open, so we had to switch nets and head for Johnstone Strait.

I was a bit nervous about it all, because I was an extra crewman. But no one seemed to mind. The skipper and I got on like a house afire. He was a radio freak and I knew some secret channels used by certain people and we scanned them constantly. A large part of the fishing industry revolves around information, and the skipper and I monitored the radios, compared notes and tuned up the sounders and electronic gear. We were a wheelhouse team.

When we did fish, I had time to figure out the skipper's style and personality. First, he was a mutterer. To no one in particular, he would say, "Just gimme the facts." Or he would comment on a snatch of information blaring over the radios, things like "Yeah, and pigs can fly." Some of it was quite revealing, such as "You have to be crazy to be a fisherman." Pause. "You have to be crazy *not* to be a fisherman."

Sometimes he sat in the skipper's chair in the wheelhouse and seemed to brood for hours. Other times he paced up and down. What surprised me was that he didn't acknowledge my presence up there with him, except when he got a thought he wanted to explore, or asked me what one of the radios was telling me. The thing he said the most often was "Turn your radio on." This seemed to apply to me, but also to be a kind of philosophy for him operating as a fisherman. He said it to the crew when he felt they weren't listening, or as a standard comment to what he heard on the many radios.

Fishing was tough. A goofy three-hour opening on the Sunday and then a twelve-hour opening on the Monday and then a twelve-hour extension on the Tuesday. We fished the lower Strait, and Tuesday morning were in the fog at Bates Pass. Lots of chart work, watching the radar and sounders constantly, and radio work to find out where we were and where everyone else was. The skipper got to say "Turn your radio on" lots of times. Working in the fog is tough, because the tension level is high and it makes you edgy. We did some smart things and we did some dumb things, such as setting behind other boats, but we constantly worked the gear, which is what fishing is all about these days.

I liked it. The crew were pros so they trained me well. They watched me doing my drumming and other chores and saw that I knew my onions, so they didn't hassle me. Except for one incident.

We stopped in Alert Bay Tuesday night because some of the crew are from there. I was sitting in the wheelhouse listening to the radios when one of the crew came in. He picked up the binoculars and watched the seine fleet head south. Recognizing my ex-husband's boat, he said to me, "Hey, there's your old man's boat, why don't you give him a call?" I think there was just a touch of malice or at least some pointed teasing in his remark.

My emotions or reactions were none of the guy's business, so I quietly said, "He doesn't have his radio turned on." Without thinking about it, I had summarized the problem between me and my husband.

Off we went to the Blue Line at San Juan, all our radios turned on. Now that I think about it, somehow it is also a good rule for living. Turn your radio on. Someone may be trying to tell you something, and I'm not just talking about fish.

Fish Wars

Just what kind of a war is a fish war, anyway? I thought it was supposed to be us against the Americans to get them to sign a salmon treaty. Last time it flared up, the plan was to put a wall of nets and troll gear out in Area 20 off the west coast of Vancouver Island and mop up all the Adams River sockeye before they passed through American waters on their way to the Fraser River. Well, it didn't work out that way. El Niño and its warm temperatures shoved all the fish through Johnstone Strait and the gap between Vancouver Island and the mainland. All those rumours about gillnetters being forced to fish in the waters off the west coast of Vancouver Island became hot air. Then everyone in the industry low-keyed it all and said there was no war; we Canadians, good guys that we are, were just going to fish as we normally did.

So we fishermen went back to our own homegrown wars, such as the one we fight every year about dividing up the catch among the various groups: Natives, sports fishermen, and the three gear types in the commercial fishery. These wars take place in offices and their artillery consists of faxes and phone calls.

Out on the grounds there was a different kind of war. We—Uncle Ed, the deckhand and me—got into it in Johnstone Strait. As trollers, we were behind in our allocation of fish and we had to go there to get our share. Ed got to use his knowledge of strong vulgar language against some gillnetters we were fishing alongside off Port Hardy. The problem was we, the trollers, could fish until midnight and there was a gillnet opening at six o'clock. Gillnetters want a "dark set" and we want the "evening bite." Both, of course, want money, as the cash price for sockeye had leaped to over three bucks a pound. When each fish is worth fifteen bucks, we're talking real money here.

These little skirmishes on the frontier were not the real fish war either. The real war comes when you, the fisherman, put your skill on the line against salmon. The preparations for battle are time-honoured and painstaking, and they start with the boat. A troller is a fishing machine and the first thing you worry about is the boat itself. Seaworthiness, maintenance problems, comfort and so on are the usual things we think about. But the big question is: is the boat "fishy"? Some boats seem to catch more fish than others and fishermen worry about why. "Getting it right" may mean changing the prop, setting up the electrical charge given off the boat (talk about voodoo stuff), redoing the shaft bearings and anything else that affects the sound of the boat in the water. There is no science to it and no books on it. Trial and error are the order of the day.

Next the boat must be rigged. This means setting up your poles and rigging so that they go out right and can be brought back in easily and at the same time catch fish.

Trollers have all sorts of special gear and tools to do this job. Getting just the right tension of springs and the right sound of the bells on the poles require a certain amount of tinkering. The same applies to the actual fishing gear. Clips, leaders, flashers, plugs, hooks and lures are an art and a science in themselves. What colour is your hootchie-kootchie? This is a valid and serious question in the war against the sockeye. Equally serious are the various bits of magic potions that some trollers put on their gear. Some are believers in sympathetic magic and put fish slime on the hooks and flashers. Others are big believers in cleaning everything in WD-40. Others believe in dishwasher fluid. Some think it is all nonsense. Ed and I belong to the WD-40 school of thought. We use it in everything but salads.

Picking your battleground is another problem. In the old days trollers went anywhere, some even venturing up to the Fairweather Grounds off the Alaska panhandle. Nowadays, with all these area closures, we are getting more and more restricted to being "inside" or "outside" trollers. One year, with no fish off Vancouver Island, we went to Johnstone Strait. The sockeye were acting crazy, so we had to have a crazy plan to catch them. Each type of salmon likes a different depth, and each one is best caught at a certain trolling speed with its own special gear setup. That, at least, is our experience. This particular year there was no lack of big schools that we could easily pick up on the sounder. The problem was catching them. We would be on fish for hours at a time with no bites. Very frustrating—we drank lots of coffee and worried about what we were doing wrong. Then suddenly, they would be on the gear.

Sometimes it was too much for Ed. He would set the gear for spring salmon. This was a no-no as we had "non-retention of springs and coho" in effect. This meant we weren't allowed to catch them. Ed said it was just to get one to eat, but I knew differently. Ed was a man at war, a war in which he would give no mercy. Besides, I think he was mad that we got nothing in Blackfish Sound, which is where he knew how to fish, and we got good catches off Masterman Island, which is where I knew how to fish.

The war took us all over the coast. It was tough but it was a fair fight. We had a good war. But there are always more wars. Every year I wonder: Will we get at least a truce for next year, never mind a real peace treaty? The struggle continues.

FALL

INTRODUCTION

Living in Equinoctial Times

Fall is a strange time for fishermen. For other people, fall is when summer weather ends and fall weather begins. It is simple, and if you still can't figure it out, you can always look at the calendar. You'll see that September 21 is marked as the fall equinox, when daylight and dark get equal time. After that the days get shorter and winter comes.

But for fishermen, two things have to happen for it to be offically fall. The first is that they have to stop fishing sockeye and start fishing coho and chum salmon. Sockeye are usually the big money-maker for salmon fishermen. The end of the sockeye signals the end of summer, and if you have missed out on sockeye you could be staring at a poor season. You will get a second chance at that particular run, but it will come four years in the future, so it may as well be an eternity.

At any rate, in the fall the focus switches to coho and chum ("dog") salmon. Coho fishing, along with the coho themselves, seems to be going the way of the dodo bird, so any wisdom I could pass on about coho fishing would be pure speculation. Seine and gillnet fishermen used to get a crack at them, but not in recent years. Now it seems only Americans and sports fishermen get to take coho. So you have to think chum salmon, which all fishermen call dogs because when they get close to their home river to spawn, they develop big jaws and savage-looking teeth that make them look like rabid dogs.

But first, Mother Nature has to do her thing. Here on the west coast, the west wind is the prevailing wind from spring to fall. In the summer, about four o'clock in the afternoon the west wind comes up and cools everyone off. But with the equinox, something happens to change the way the wind blows. The prevailing wind shifts around the compass and becomes a southeaster, the winter prevailing wind.

However, the calendar is one thing; the wind is quite another. In a better organized world than this one, the wind would shift from westerly to southeasterly at twelve noon on September 21, but it doesn't happen that way. In short, during equinoctial times—I checked my Oxford dictionary and yes, that's how you spell it—the wind can't make up its mind. For gosh sakes, a southeaster can come up out of nowhere in August and blow all the sockeye out of Juan de Fuca Strait. Similarly, in October a westerly can spring up, confusing the poor mariner. Put it all down to equinoctial weather. It's a time of transition and can be confusing.

It is a time of change in weather systems and it causes

me to reflect on the changes in my life and in the fishing industry. First of all, fishing used to be fun. It was your living and it was serious business, but it also had great opportunities for fun. In the fall the trollers would be finishing up the last of the coho. Maybe some of them would convert to gillnetting to get some dog salmon. The seiners would be moving south. Above Cape Caution, the summer dogs were all caught and the fall dogs were eagerly anticipated. In Johnstone Strait the seiners and gillnetters awaited the silver brights—dog salmon that still looked like sockeye, but were usually twice the size. Nice-looking fish that fetched a nice price. Not only do dog salmon change colour and appearance as they get close to their home river, they also spawn in the lower part of the river. Not for them the long journey hundreds of miles up a river to some mountain lake. The river mouth will do for them. Maybe that's why they don't bother to come until late in the year.

The dog salmon runs usually peaked around October 1 in the strait, and if you were lucky enough to get lots of fishing time it was fun. In the upper strait the schools were smaller—and wilder, it seems to me. They had a propensity to do odd things like go the wrong way. You would estimate they were going south to the river on the flood tide and as soon as you set your net the rascals would go north. So you made lots of weary sets as you travelled around looking for the mother lode.

In the lower strait schools were larger, but harder to find. And as the fall progressed, there was always the southeast wind. The wind affected the fish. It could blow for days, impeding the fish and schooling them up. Sometimes we got to fish dogs off the west coast of Vancouver Island and that made us very aware of the southeast wind. The problem was that it was never predictable. A small wind could turn into a gale or a gale could blow for days and then out of nowhere would come a flat calm and sunny skies and you would think it was summer.

But for all this, fishing dog salmon could be fun. In a good year, they were kind of a bonus after the summer money was made. Now the dog salmon don't seem to come as much any more. Prices have fallen because the Japanese have figured out how to restore the big runs of their own dogs. Fall fishing has become a lean and bony business, with the coho gone and now very few dogs.

Because fishing is a curious occupation it has attracted all sorts of people. Ex-prairie wheat farmers, ex-computer technicians, ex-(until the winter at least)journeyman carpenters; the one thing they all have in common is the burning desire not to have a steady job. There are no wages in fishing. Everything depends on the amount of fish you catch. It takes a certain kind of person to function under those conditions. You would think that commercial fishermen would be stressed out, gloomy, cynical types. In reality, in the main, fishermen tend to be philosophical, easygoing, and known for a certain camaraderie. What this means is that you can have fun with other people while fishing. You can share private jokes at other fishermen's expense, fish together, help each other out of serious jams, and even blockade Alaskan ferries together. The beauty of it all is that once you're back home in port, you don't have to see all these people again.

Good times with the west wind, tougher times with the southeasters. Fishing still has all the great things that drew me to it in the first place, but like equinoctial weather, times are changing. There's a really mean southeaster blowing, and in this case it's caused by just too many people trying to make a living from the fish. So gear types fight with other gear

Living in Equinoctial Times

types, sports fishermen are always mad at commercial fishermen, the north is pitted against the south and on and on.

The fall equinox is a time of change and it makes me think of change in the industry. Times are tougher. Fewer fish and lower prices have hurt us all. It is a hard decision to give up a way of life that is the only one some people have ever known. Because it is a different sort of life it is difficult for outsiders to understand what we have lost and what we want for the future. Some want to turn back the clock but you can't. Out of frustration, some want others to give up their share of the fish. Equally, some have survived and are doing quite well, thank you very much. People who could manage their debts and kept abreast of the technology and had the right attitude to change seem to be okay, but I don't really know. And so the debate rages.

I guess the fall with its equinoctial weather tells me that change is always with us. I've lived through several El Niños (the warm Peru currents) which cause salmon to do strange things. I've seen the coho slowly disappear, and dog salmon fishing reduced to twelve hours for the whole fleet. I've seen myself get older and not so eager to try new things. I have mastered all sorts of new technology—from plotters, which draw a line on a computer screen so you see where you have been, to a satellite locator which tells you where you are and how long it will take you to get where you are going. I've even tried my hand at sonar, which allows you to look around the boat under water and see where the fish are exactly. Radios, sounders and temperature gauges are now a normal part of what I can run on a boat.

But technology doesn't help the human soul. These are changing times and I must get ready. It's easy to prepare a boat for bad weather—batten down the hatches, secure everything, lower the boom, close the portholes—everyone knows how to this. For the turbulence caused by changes in the industry I am not so well prepared. Well okay, I know what to do, but the trouble is it is so boring. Don't get in debt, don't spend, collect all bills owed and make the accountant's goal your goal: a positive cash flow. The direction of my life has changed. Before, I fished and I got money to live. Now I have to plan out my life so that I will have money so I can go fishing. I guess it is a good thing that equinoctial weather is part of fishing, because it has trained us to get by in equinoctial times.

The Secret Life of Dog Salmon

You watch. Then, out of the corner of your eye, a flash of white. You watch there more closely now. In less than a minute, but it seems longer you are so intent, you are rewarded with several quick splashes of white, one right after the other. You smile and say quietly, "dog salmon."

We were in Johnstone Strait again, after "dogs"—or *Oncorhynchus keta* to give them their scientific name. This was a few years back—around the time we had that big constitutional referendum. Although it had been a poor year for sockeye salmon, there sure were a lot of dogs and we were after them.

Fishing for salmon is different wherever you go and the different species require different skills and knowledge. We catch dog salmon all during the year but they are incidental until about October, when they return in great numbers to spawn in the mouths of rivers. So here it was mid-October and the big southeasters were starting to blow, making our lives more difficult, and we were thinking seriously about dog salmon.

They are big fish, twice the size of a sockeye and three times the size of a lowly pink or humpy. Also, they act different—sort of like sockeye, and they jump the same way, but they have their own way of travelling and acting as they go through the strait. We had endless discussions about how to outwit them—where to set to catch dogs and how to fish them. I told the rest of the crew that we needed to know the secret life of dog salmon, then we could catch them.

Of course everyone laughed, but I guess after being anchored up for so long waiting for an opening it was the only thing we had to talk about. As we talked I made lots of notes about the various subspecies and how they were in the old days—which ranged from about four years ago for the youngest of the crew to about forty years ago from the skipper's viewpoint. Some species are only about seven pounds and some about fifteen. For the fish buyers they split them into "summer dogs" in the north and "brights" from the strait. The buyers want the silver brights and not the ones that get darker as they get closer to spawning.

I kept my notes in a folder in the galley with a label "The secret life of O. Keta." Somebody wrote in it, "We have a secret life also." So at mealtime I asked everyone to tell me their secrets. Everyone said either they had none or if they told it wouldn't be a secret any more.

So I got to thinking about the secrets we do have. We have them and don't reveal them except, it seems, to strangers. For example, I couldn't get anyone to tell me how they were going to vote in the referendum—except Pete, the beach man who ties up the end of the net. Pete said he would vote Yes because he had his reasons. I pushed him to explain but he replied it was his secret and I could mind my own business.

We were fishing and we took whatever we could get, silver brights or otherwise. There were too many boats and we had to set blind offshore sometimes just to compete. We had been working in the lower strait as the dogs seem to school up better there, and the old pros had got some good catches. The skipper kept saying "the fat lady is singing" by which he meant that this was our last chance to make any money for the season. The tides were lousy, the wind was blowing, too many boats and strange grounds. A poor combination but then one day we started off with a couple of good rampers, meaning we had to use our stern ramp to bring the fish aboard, a sign of a good catch. After lunch we moved farther down and looked for a good opportunity to set.

The lower strait is famous for strong tides and we were able to slip into a nice spot and set. Our new beachline was tight as a violin string as we tried to close the set. Then I saw Pete and Tom scrambling about on shore. I thought the beachline had snapped and that we had lost the net's connection with the beach. Then Pete got it tied up again and we were able to close the net without any problem.

When we finished the set the skipper yelled at Pete, "Why the hell did you let go?" What had happened was Pete had let go without a signal from the skipper. This was next to mutiny. Pete explained that he thought the beachline would break, so he let go and moved down the beach and tied the net up again. He said he had done this very thing before at this spot before. This we all felt was a barefaced lie. The skipper was still mad and continued to berate Pete. Pete finally looked at the skipper and said with fierceness, "Don't you be unconstitutional with me!"

There was a stunned silence, then within a few minutes we all were back at work. Like the rest of us, the skipper had no idea what Pete meant but the matter was considered closed.

We got some fish, salmon season ended, the World Series ended and the referendum also ended. The secret life of dog salmon continued, although, like the thoughts of Pete the beach man on the Canadian constitution, it was also still a secret.

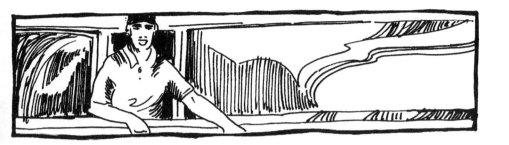

The Man Who Remembered Everything

Recently, I got to meet a guy I had heard about for a long time: the Man Who Remembered Everything. I was interested because I have this thing about knowledge in the fishing industry. There are no books on how to fish. The only books that are any help are the tide book and maybe *Captain Lillie's Coast Guide*, but I know guys who don't read either one.

Because of this, fishing remains a kind of "monkey see, monkey do" proposition. If you want to know how to seine in Hakai Pass you watch the fleet and do what they do. Same with trolling or gillnetting. However, you do have to have some knowledge, skill or experience to make it all work for you. If there was a how-to book on gillnetting, its chapters would have to have names like "What to do when your net is rolled up and the tide is pushing you onto the rocks." So it is important to get as much knowledge as you can, about everything. And you have to remember it all. A good memory, even if it is a short memory, is a requirement.

Notes from the Netshed

I had a funny experience when I first crewed on a seine boat. On my husband's and my boat I did all the bookkeeping, so I got into the habit of keeping lists of everything, putting all paper into the proper envelopes, and making sure we had all the manuals and books. This habit carried over when I crewed on the seine boat, so that even though I was not a part-owner, but just another wage slave, I still wrote things down.

The skipper was highly amused. He told everyone that I was a real case because I had been to university and was educated but couldn't remember a damn thing. He thought a poor memory was a sign of mental retardation, and he began reciting the feats of someone he considered truly intelligent, in fact brilliant. That was the first time I heard about the Man Who Remembered Everything.

Apparently he was noted for this skill even as a kid. He also was famous for one other thing. He was able to tell you the time almost exactly, even though he never carried a watch. To me he was important; to many he was a colossal bore.

Now, I have been around long enough to take claims like these with a grain of salt. In the first place, what everybody knows for a fact is not always what is true. And in the second place, I suspected that in the end, what the Man Who Remembered Everything knew was not much. So what if an elephant has a good memory: How important is the stuff that it has to remember?

In general fishermen are a nondescript-looking bunch. Each one looks like the rest and you cannot tell the rich ones from the poor ones, a seiner from a gillnetter or whatever. So when I met the Man Who Remembered Everything I did not see anything about him that would set him apart as someone special.

The Man Who Remembered Everything

Uncle Ed and I were in a store in Steveston buying one of those heat lamp setups that you put in your boat in the winter to stop moisture. Ed saw the Man Who Remembered Everything first and of course said: "Remember me?"

Well, of course he did, and he knew that Ed was giving him a little shot directed at his skill, so he shot back at Ed: "Still robbing blind men?"

Ed got red in the face. Obviously this was something that was true and that Ed was not proud of—something that happened a long time ago, as I had never heard of it. Jeez, robbing a blind man, how and why had Ed done that? Somehow I knew that I was never going to hear of it, at least not from Ed. Ed introduced me, and the Man Who Remembered Everything recounted a long anecdote about my father, which did not impress me because among fishermen it was common knowledge. Nonetheless I wanted to at least see and hear something of this living legend. So we went and had coffee just to talk over things and get industry gossip and pass the time. This is a necessary task many fishermen indulge in.

The Man Who Remembered Everything owned a beat-up seine boat and so talked seining to us. We talked trolling and gillnetting. We cursed the scientists and bureaucrats, slandered close and distant friends, talked about satellite navigation, fish prices and so on. None of this was out of the ordinary. I had heard nothing to impress me about the memory of the Man Who Remembered Everything. While he and Ed talked, I mused about memory. What should you remember? Where you left your car keys? How to set in Double Bay? The name of your bank manager? The colour of the linoleum in your elementary school hallways? (In my case it was brown.) As I understand it, theoretically this is all in there and it is up to you to get it out.

167

Then, like an engine revving up to full power, the Man Who Remembered Everything started doing it. When the name Nitinat came up, he told us about a certain Matt Niklich and how the *Mary N* had rolled over in the pass, long ago. He knew the crew's names and a detail or two about them all. Matt had made thirteen sets for 39,000 fish in the Charlottes and was going to finish off in Nitinat. Then he told of the *Roy Roberts* going aground in Skidegate Channel, then of the creek sockeye in Fife Sound (three pounds, in early June), and so it went, on and on. One thing led to another, but because he really knew his stuff it came out all in one piece, and he came across as interesting but not garrulous. It was the kind of conversation fishermen have that sounds like a bunch of guys shooting the breeze, but it is actually full of information and asides and insights and conclusions. It may be BS, but it's BS for the record. I could see that he really didn't remember everything but had made an effort to sort things out and put them in logical order, and had worked at perfecting his ability to recall the information. I guessed that he also consciously blocked out a lot of stuff that either wasn't relevant or just had two many little details to fit in his memory storehouse. He could remember, but he had to be selective in what to remember.

Back at Ed's, I talked about how I was impressed with the Man Who Remembered Everything. Ed claimed not to be. He noted that lots of fishermen knew and remembered lots of things. Ed cited examples of his own skill, but I pointed out aspects in which the guy's memory, even compared to Ed, was incredible. Perhaps haunted by the incident of the blind man, Ed finally agreed.

My next comment was to wonder why the memory expert was not rich. Ed snorted: "He don't know what he knows, he just remembers it." Ed has a point there, but what it is I've forgotten.

Fisherman's Wife

I said good riddance to the end of last year's fishing. We went to Nitinat for a dog salmon opening, just about got shipwrecked in a blow that came up in the middle of the night, and in the end there were still not enough fish, which means a tough winter was coming up for me. I even thought about going up to Double Bay and joining the live cod fishermen, because I had to get some money somehow.

To find out the industry gossip, I did the rounds of all my pals. Fishermen's wives are good for news, so I phoned them all. It would be great to get a net loft job for the winter, but to get that you have to do a lot of other stuff first, such as find out who is thinking of building a herring seine, or whatever.

I also ended up phoning Uncle Ed. I hadn't been seeing too much of Ed since I got back from fishing, because he had a new wife. She and I didn't necessarily see eye to eye. Ed met her when he was doing some work for—can you believe it?—an old folks' home. She was some sort of volunteer from a church, her husband had died a while ago, Ed came on the scene and so they got married.

I was surprised as hell, because Ed's first wife left him many years ago. Since then, he had several of what are now called "meaningful relationships," but basically the bachelor life seemed to suit him. None of my business, but that's how it was.

Anyway, when I first met Ed's new wife, I checked her out, and while she was as pleasant as pie to me, I noticed that her mouth smiled but her eyes didn't. In short, she was a firecracker of a wife. My first mistake was to say that it was always a pleasure to meet a fisherman's wife.

"I'm Mrs. Smith," she said, "not a fisherman's wife." I got the message.

Now, fishermen's wives come in all shapes and sizes, like their husbands. But they are not the same as their husbands. While I could argue with Ed and point out his errors or whatever, his wife obviously was not to be trifled with. So I steered clear of her and was suitably respectful to Ed.

We talked on the phone a while and I expressed some reservations about the coming winter and so on, and Ed asked me over. Of course I had to go, but I was nervous as hell. We sat in the kitchen of their townhouse, and Ed's wife watched us like a hawk as Ed and I went through my financial problems. I needed his advice, because for all his rough edges he is good at getting to the root of problems and solving them. His wife, on the other hand, made me nervous. I think she thought I was trying to borrow money. In spite of her listening, I explained my problem to Ed in detail.

My problem was that lack of money continued to roll in. The herring money was all gone as I had paid my debts, but the salmon money was not enough for the winter. Plus I had a loan at the bank that was coming due, and no money to pay for it. I could see Ed's wife narrowing her eyes at this, but she said nothing.

Ed and I talked for a while about potential net loft jobs, the possibility of going dragging or something in the fishing industry. But it was just talk. After a long silence, Ed's wife asked who my banker was.

"Oh," she said, "that's the church's bank." Looking me straight in the eye, she asked if I would be happy just getting a new loan renegotiated so that I could let the interest build up and pay it next year. I agreed that this would be okay, but I doubted my banker would go for it.

"Let's find out," she said. Off to the bank we went.

The manager certainly knew Ed's wife, giving her the smile bankers give to customers who have large accounts and never bounce a cheque. He was surprised to see us together, as he knew about my unsuccessful gift shop and my occupation as fisherwoman.

"This is my niece," said Ed's wife, "and we are here to refinance her loan on terms that can assist her." Already the bank manager was doing some sort of calculations in his head, but she pressed on. "My husband and I have always found this bank to be helpful and competent, and the church, as well as many of its members, bank here because you are so helpful."

The banker looked like a red snapper that had just come up from fifty fathoms. Ed's wife was saying that if he didn't do right by me and my loan, she would transfer lots of money elsewhere. Smiling, but in great pain, the manager said he had few fishermen's accounts and, while they were certainly different, he would nevertheless, in this one special case, bend the rules to accommodate me.

"Today?" asked Ed's wife.

"Today," he replied. And so it was done. We went into

his office and he wrote it all up, and I signed it. I was only postponing the payments, but my small pile of cash was preserved. We came out of his office and said our goodbyes.

"The bank values your business and that of the church," said the manager, "and you certainly have a great appreciation of the fine points of finance."

Ed's wife gave a small smile and said, "Oh, I don't know about that. You see, I'm just a fisherman's wife."

High-Test Beer and High-Tech Fishing

All the old trouble about beer broke out again last fall. The trouble starts because everyone thinks he is an expert on beer just because he drinks it. So arguments about the quality of beer and which one is stronger become nasty almost immediately.

Of course, everyone has their own opinion about different ways of measuring alcohol percentages, but no one has any concrete proof, so they try to make their case by shouting. And each person has his own individual ideas about the ideal conditions for drinking beer: after a hangover, before supper, as a chaser, while drinking muscatel wine, and so on. Beer arguments are always the same, but they are always fresh and interesting to the participants.

The beer arguments started up again around here because I thought I'd try my hand at brewing some beer of my own. Uncle Ed sampled it and said what he always says about homemade beer: "Put it back in the horse." This raises

High-Test Beer and High-Tech Fishing

an interesting question: how do you find out if homemade beer is any good?

You have probably been in this situation yourself. You are in a home-brewer's house. He offers you some of his beer. You drink it. Your host wants to know how it is. Not wanting to hurt his feelings, you assure him that it is of the highest quality, but he persists. In a fit of niceness, you tell him it tastes better than any of that stuff they sell at the wine and beer store. He beams. And, unfortunately, continues to make beer that he cannot get rid of because it tastes like something that will clean the bugs off your windshield.

Anyhow, with all the new beers on the market a new word was bound to spring up. People here began to call the new-type beers "light," and call the old type "high-test." I think they called it that because high-test gasoline is supposed to have a higher octane rating. So in this time of changing beers, those who stick to the old ways continue to ask for high-test. Uncle Ed is a high-tester. He drinks a beer called Old Style. "Old Style makes me hostile," he says when he cranks open another bottle. Ed is strongly against both the new beers and the home-brewers.

Around that time, the banks had gotten worried about the state of the fishing industry and decided to do something about it. Which means, in bankers' terms, that they repossessed quite a few boats, mostly trollers. Ed and I were going around the country looking them all over and hoping to come across a bargain that we couldn't pass up.

Ed had recently been down to the States and on the way back loaded up his car with cheap American canned beer. So wherever we went around Vancouver and over on the Island, we broke out the American beer. Of course, the beer arguments started right away. Ed drank the American beer, but complained bitterly about it. He wanted his Old Style, but he can't resist a bargain.

As we inspected each boat, Ed looked at the engine and electronics very closely. He is a great believer in high technology, and he keeps up on all the latest developments. If we were to buy any boat, it was going to be a high-tech boat. All the time he looked over these boats, Ed was drinking American beer and complaining about it. This seemed like a good reason to start an argument with him. I pointed out the inconsistency in his attitude. When it comes to beer, he is conservative and conventional; but when it comes to electronics, he believes in "out with the old, in with the new." Ed rose to the bait, and tried every way he could think of to deny that he was inconsistent. Ed hates to lose. I knew I had him when he started his old nonsense about women not knowing anything about fishing. Then he started stalling. I knew it was so he could work on his argument until he somehow got it the way he wanted it.

I would leave him alone for a day or so and then I would start the argument all up again, to see if he had anything new to say. Finally the answer came to him. I didn't think it was much, but it suited him.

Ed said he wasn't inconsistent; it was just that in any field of endeavour, he insisted on only the best. "My beer has to be the best, and the electronics in my boat also have to be the best. So you're wrong again," he concluded triumphantly. That ended the argument and any further discussion.

We finished the last six-pack of the American beer coming over on the ferry from Nanaimo. I was glad to see the last of it and so was Ed. Whether it's beer or boats, chasing bargains can be exhausting. I doubt if either of us will be bringing up the old arguments about beer for a while—at least until we run into another home-brewer.

Staying Alive

I always come out of a poor salmon year with one obsession burning in my brain: staying alive. Not depressed, just determined, because there is little I can do to change the situation. As I've heard that misery loves company, I start to look around for other poor souls like me. Unfortunately, there are few single (well, "de-married" as a friend puts it) women I know who work as fishermen. I use the word "fisherman" to describe what I do although someone, somewhere, has decided to call us "fishers" to make it gender-neutral. No one I know calls himself a "fisher" so the newspapers are being silly, but that is normal, I guess.

Speaking of misery, last time it happened I thought I would go and see Uncle Ed. Some trollers had a bad season and I know Ed wasn't as spry and aggressive as he used to be, so I figured he would make good company for me as I catalogued my woes. The trouble with Ed is you never know which way he will jump. By that, I mean he is so darn contrary that there is no way you can predict what he will say or think about a particular problem. While he may irritate you, it also makes him an interesting person to be around, if you can stand the heat. To get ready for a morning with him I went to the Net Shed Cafe in Steveston and had the high-cholesterol breakfast of bacon and eggs, hash browns, with toast and coffee. A traditional fisherman's breakfast, I thought. No one I knew in there but I saw the usual tables of gillnetters telling each other lies about great drifts they had made in the north.

As usual, Ed and I sat in the kitchen while his wife fluttered about pretending to do things but really listening to every word. Right away Ed started to surprise me. I moaned that I might not have enough stamps to get unemployment insurance. Ed said he didn't apply for it and never would. Surprise One. Surprise Two was his reason. "How can I complain about the handouts to Natives if I draw pogey?" he demanded. Thinking he might be anti-Native I asked about the recent problem of the missing sockeye in the Fraser River that some Natives have been accused of catching. "Those fish were never there," said Ed.

Now I know better than to get into any sort of argument with Ed, but I asked him how the hell he knew that. Ed explained that during small runs there are lots of five-year fish that have come a year late, just as big runs have three-year fish that come a year early. So the fish will come next year. Now Ed is no biologist and never expressed any opinions about these sorts of things, so I was puzzled as to why he knew this stuff, right or wrong.

Of course later I found out that one of his ex-deckhands (and there are many) works as a biologist and phones Ed from time to time. So Ed was up on the latest biology stuff, or so he maintained. This talk was leading me nowhere and I wanted to switch the topic to my problems when the phone rang.

Staying Alive

It was one of Ed's pals who phones him to keep him up to date on who's dying, who's doing poorly, and Canucks hockey, Ed being a big fan.

Ed hung up after giving his summary to his pal. "The trick is to stay alive, stay alive," he said. I suddenly realized that without knowing it, Ed was giving me the answer to my problems.

Staying alive in this day and age as a fisherman requires a bit more than in the old days. By tradition I should have been smoking dog salmon for the winter. My pantry should already have been full of jars of sockeye as well as blackberry and salal berry jam and so on. Wood should have been cut and cedar kindling made. I guess all that still takes place upcoast but here in the city it sure is different. To stay alive you need a good credit rating, positive cash flow and liquid assets. None of these do I have. And I certainly won't and can't go back to the good old days which in my opinion usually weren't. These are my good old days.

I wanted to leave Ed's as soon as I figured out what I was going to do but having unloaded on him all my worries I had to listen to his catalogue of woes. I listened respectfully but I really didn't care, since personally I have no prospect of owning a troller again, at least for some time.

This was ungrateful of me but I knew Ed would be all right. He had his plan for staying alive and it was up to me to get out there and get working on mine. Staying alive meant physical health, of course (Ed had quit smoking, hadn't he?), it meant economic health (that cash-flow stuff) but most of all it meant mental health. By that I mean if I kept up the right attitude and did some planning, the lack of fish and fishing time would be no big deal.

I know all this sounds like whistling past the graveyard. A couple of weeks more fishing time would have meant a little more cash (okay, that's a big Maybe) and enough stamps for UI, never mind what Ed said. But I knew if I kept my chin up and looked around, maybe I'd find a job on a dragger or in a net loft or something, anything.

Meanwhile, I said goodbye to Ed and went off to get a case of Kraft dinner for the winter. It wasn't smoked dog salmon but I was staying alive. To stay alive you need food not only for the body but also for the soul. At the end of the season, thanks to Ed, I think I have both.

High Jinks on the High Seas

For a few weeks last fall the weather was depressing and so was I. I know that if you look at maps of the BC coast it says "temperate rain forest" but this was ridiculous. To top it all off I had to go to the bank. The manager phoned to say he wanted to "update his data on my file." This means because I'm a single woman and a fisherman, he wanted to run his eagle accountant's eye over my file, looking for an excuse to make my life more difficult than it already is. I have no assets other than a beat-up car, a computer, books and clothes. But I do have a herring punt licence and I have always paid my bills. The current crisis in the fishing industry had put the bank on its guard against those of us who might sink without a trace into financial limbo.

For some reason I got steamed up about it all. They have more information on me than I do on myself. For a small overdraft protection, a line of credit and a small loan they were asking me to predict my future ability to be financially solvent. If I was that smart I would be rich.

To get in a better mood I went over to Uncle Ed's house. This can be a gamble. Sometimes Ed is "in one of his moods," as his wife puts it. This means Ed is barely communicating, or he is ranting and raving about some new idiocy the world has inflicted upon us working people. At other times he can tickle my funnybone. Ed, believe it or not, has a few interesting talents. When he is in good spirits he can sing in a pleasant voice. In fact, it is a high, sweet soprano like a girl. Also he knows *The Cremation of Sam McGee* all the way through, and my favourite, not heard much nowadays, *The Face on the Bar-room Floor*. Ed is also a storyteller *par excellence*.

Hoping to catch Ed in one of these moods I gave my secret knock (three and two raps). Ed was there with one of his pals. I knew it would be an okay time as Ed's pal was a retired troller (is there any other kind?).

They were playing the "remember the time you ..." game. Their topic was goofy tricks and the things they had done when the world was young and fishing was simpler and better. I had forgotten the things we did to each other when we had time on our hands when we were fishing. I am talking about the good old days when fishing was fun. Maybe some of the things were foolish—such as tying a rope to the foot of a sleeping person and waking them up and watching them fall over when the line tightened up. Much safer was putting an inch of water in their gumboots and watching their face as they put their feet in them. A favourite of mine was to use that very good glue and glue up the pantlegs of a pair of pants so that the victim could not put his leg into them upon arising from a good night's sleep.

This all may sound juvenile, but a big factor was the personalities involved. A good victim was one with absolutely no sense of humour, or maybe even one who would go ballistic with rage. Even more fun was a victim who would bide his or her time and retaliate in a suitable manner. Sometimes, like the one Ed and his pal were talking about, the pranks were cruel, and deliberately selected to play upon some weakness or failing of the victim. Joe (not his real name) was hard of hearing and also was a radio hog. He was always on the air, loudly wondering how his radio was "putting out."

The first part of Ed and his pal's stunt was to get the guys in their group—they were all trollers—to agree not to respond to any calls from Joe. They pretended they had never heard him. This drove him nuts and he spent good money on hiring radio technicians to fix something that was already okay. The next part of the stunt was even worse. Again, nobody replied for a week or so, and when they finally were in touch with Joe they told him they had heard his radio and they had replied. Now the victim had to get his hearing aid tested. Cruel, but great fun for those in the know.

Since I knew some of the actors in the various dramas they recalled, I spent a pleasant hour or so with Ed and his pal. I told Ed that his stories could be sold to a magazine as "high jinks on the high seas."

"Maybe so," said Ed, "but all the high jinks are now on shore." He gave me that look of his that means he doesn't like something. I suspected it was the current state of the fishing industry. If only we could spend more time out on the water—maybe then there would be more time for high jinks on the high seas.

Preventive Maintenance

Handy Andy is no more. That is to say he is dead. While his name was not really Andy, he was the village handyman. Nowadays there are no more handymen. In simpler times they were the nonspecialist or jack-of-all-trades that had a specific role in the various trades that are necessary for a village or a community to function. They could, and did, fix anything, but their stock in trade was their willingness to tackle any task and somehow get it done. Handy Andy was such a person. He cleaned out septic tanks, did a bit of electrical wiring, knew how to fix a leaky roof ("The leak is never where the water is," he once told me), could install a toilet and knew much about pumps and generators. In many tasks he would defer to the specialist. Digging and manual labour were not beneath him but it had to be high-calibre digging, otherwise he would suggest the village labourer as a more suitable person for the job. He had a trade and it was as handyman.

When I heard of his death I knew that it signified the

Preventive Maintenance

passing of a way of life. In those days not so long ago you had things that could be fixed with a minimum of knowledge and tools, and so a handyman was a logical person for a village to have. Now, nothing gets fixed. It is replaced, as anyone who owns a toaster knows. Cars need computers to tune them up and outboards can only be fixed by rocket scientists. Who amongst us can program their VCR, let alone fix it? I am sure as Andy got older there were fewer and fewer things he could fix and his jobs became more and more like those of a tradesman or a labourer. Maybe his last days were sad and he became less important.

In a curious way all fishermen are handymen. The list of skills they need would dazzle a vocational school instructor. On a boat, things are always breaking down or getting out of tune. There are no service stations out on the grounds so things must get fixed one way or another. Tuning in radios, sounders and sonars are small things. Replacing a hydraulic hose and those goofy fittings is another chore. Adjusting the float valve on the stove fuel carburetor is a much-valued skill. The use of Allen wrenches, torque wrenches, socket wrenches (metric and SAE) and the knuckle-busting crescent wrench is standard. Occasionally, a tap-and-die set is needed.

In the electrical area knowledge about circuits and fuses and loads is required. Boats have different voltage systems and what's what (or "What's Watt") means lots of different light bulbs and something called an inverter which allows you to run a TV set on a boat. And of course tape, especially duct tape. If all else fails, bring out the duct tape. With it you can fix your rain gear, tape a sore ankle, fix a leak, smooth over a splice and tape the water kettle down in rough weather. All kinds of other things are considered essential. Everyone has their favourites.

All these tools, washers, bolts, screws and tubes full of mysterious goop are no good without the knowledge to use them in the right place. But all this stuff is after the fact of a breakdown. What is needed is a philosophy to prevent breakdowns. This is what I remember about Handy Andy. Whenever he was called in as a consultant to a job, he discussed the problem, his fee (cash at the end of the month) and the cure for the problem. Like a good doctor he could treat the symptom but he also had a cure for the disease. It was always the same: preventive maintenance. He believed you should anticipate problems and do the little things necessary to prevent a full-blown disaster.

You will see the two schools of thought on fishboats. Some skippers and engineers let things go until the ultimate breakdown and then apply their skill. Others, perhaps more thoughtful, do the preventive maintenance. They can somehow sense when the alternator is ready to pack it up and put in a new one. They also know that pumps wear out and need surgery or replacement. Some, dreading the burst hydraulic hose or blown pump in heavy fishing will tear out everything hydraulic and replace it. The preventive maintenance guys are my kind of people. The Zen of outboard motors is beyond me so I know that every season a new ignition system is the best medicine.

As usual I try to apply these things I know from the fishing racket to the other life I live on shore. I sometimes get into silly mind discussions with myself trying to figure out if I am doing preventive maintenance in my life. You know, doing the little things such as changing my eating habits, less drinking, avoiding certain people, and in general tuning up my mind and body as well as my life. Trouble is, on a boat it is easy to order a new part, in life it is difficult to order up a new way of thinking or new parts for a damaged soul. There are few handymen for the human spirit. All we can do is place our hope in preventive maintenance.

The Fisherman's Shed

When I was married, my husband and I lived in a number of houses. We also had a number of boats and fished a variety of ways. There is one thing about our marriage that I miss—in fact maybe it's the only thing. The last house we lived in had a shed that contained all our junk from fishing. I guess fishermen's sheds are more or less the same everywhere, but I miss ours. Lately I've been thinking about just what a fisherman's shed is good for.

Ed says that a shed is just a junk collector, so he won't have one. That means he stores his junk all over the Lower Mainland with other people. But I think that if you have your own house and your own boat, it is more honest to recognize that you need your own fisherman's shed.

Some people are able to keep all their stuff in their basement but we lived in the Zuider Zee of Canada, Richmond. Also known as Lulu Island. Lulu was the girlfriend of an early BC politician, so he named the island after her. As it became built up it came to be called

The Fisherman's Shed

Richmond, which is supposed to be more respectable, I guess.

Richmond is about four feet below the high-water level of the Fraser River estuary, so it has a system of dykes and drainage ditches. A basement is out of the question, unless you want an indoor saltwater swimming pool. Therefore if you have lots of stuff, you need a shed.

You never know, the taxman may come by someday, so I had a list of everything that was in our shed. It was tacked to the inside of the door, but somehow it got rained on and was rendered useless. I was in the shed one February, checking our herring gillnets for the herring roe fishery in March, and suddenly I got frustrated that I no longer had a list of everything.

So I just had to tear that shed apart and make a list. The list was one thing, but the reasons that I had all that stuff in that shed were something else. For instance, what the hell were we doing with a shrimp beam trawl? The damn steel beam was twenty feet long and a real pain to store. I remembered trying beam trawling for shrimp, but that was years before. It's okay to have lots of gear, but twenty longline skates of halibut gear and anchors and buoy lines (some with lights) was a bit much, especially since we moved that stuff to three different places over the years.

Then there was a big box labelled "Plastic Buoys," but it was really filled with old-time seine cork floats. We had about five hundred of them. Some fishermen make ashtrays out of them, and it seems to me I had something useful in mind for them, but now I can't remember what it was. So it goes.

The stuff in the shed was in the strange nameless land that things enter when you don't have any immediate use for them, but they are not yet totally useless. Some of it could have been sold, but we would not have gotten a price that would have been anywhere near what it was worth. There was a guy out in Surrey who would take it all, but only for free. He never seemed to sell anything, so maybe his shed was bigger than ours.

I had sold all our salmon gillnets when we sold the first *Fantasy Isle*, but I kept a box of our secret plugs and lures from trolling. We had electronic gear in the shed too. When everyone had Citizens' Band radio we went through several sets of CBs, but after the craze passed our CB just sat out in the shed, its electrically powered retractable antenna permanently retracted. A depth sounder, old but in good working order, was there to keep the CB company.

Not all our stuff was fishing gear. We had a wheelbarrow that we never used but was on hand for the neighbours to borrow. That was a curious thing. How come people will borrow power saws and wheelbarrows, but nobody borrows fishing gear? Every time fishermen need something, they go and buy it. So everyone's sheds were full of stuff like ours. I'll bet if we had ever put it all in one place it would have been astounding.

In the old days fishermen didn't need as much gear as they do now, and they could always store it at the fish company's net loft or cannery. Nowadays there is a hefty charge for that service, so the number of sheds is on the increase. I miss our old shed, but it was a VW bug compared to some people's sheds, which are Cadillacs.

Ed felt like a beer one Sunday and some company so he said: "Let's go to Dwayne's." We headed out to Dwayne's place, which is out in Ladner along the river and I was admiring all the sheds. Those guys on the dyke have sheds that you could park a small airplane in. They are not just for gear, either. Dwayne's shed is unbelievable. It has the required ten-fathom stretches for making a seine marked off on the floor. It also has a table saw, a planer and everything else you need for woodworking projects big and small. But what it also has that makes it a fisherman's dream are two fridges full of cool beer. When we got there we joined six other guys who were all sitting around having a beer.

"Help yourself, Ed, and get a beer for your long-haired partner too," said Dwayne. We stayed a couple of hours. "Not a bad little shed," said Ed when we left. I didn't have the heart to remind him he was dead set against fishermen's sheds.

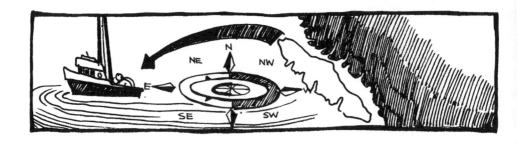

Kiss Me Hardy, For I Am Dying

Being a pack rat like me has its advantages. While reading some of my old journals I have been known to burst out laughing even though there is no one around. I don't think this is unusual.

Anyway, I had a good laugh at my former husband the other day. I was reading my notes on the names of fishing spots on the BC coast. I have always been interested in the names of islands, coves, points, lighthouses and particularly fishing spots. So I kept a notebook and while rereading I thought of when I was married and we owned the *Fantasy Isle*. Every time we were on our way into or out of Port Hardy I used to say to my husband, "Kiss me Hardy, for I am dying." He never knew what the hell I was talking about (then or any other time) and most fishermen would not either. So I had a laugh on him when I saw "Hardy Bay" in my notes.

Hardy Bay was named after Sir Thomas Masterman Hardy who held Lord Nelson in his arms as he lay dying from his wounds at the battle of Trafalgar in 1805. Nelson's last

Kiss Me Hardy, For I Am Dying

words were "Kiss me Hardy, for I am dying." In those days men kissed—and why not, for they were heroes. The geographers who named much of BC had a sense of history and blessed every rock and point and cove with a name in honour of one of their pals or heroes or, when they ran out of names, just whatever they fancied. You can look it all up in Captain John T. Walbran's *British Columbia Coast Names: Their Origin and History*, which most of us refer to simply as "Walbran," the implication being if you don't know who he is, too bad. That is what I used to say to my husband, just to let him know that I knew more about the BC coast than he did. This I guess is one reason that we are not married any more.

But I have my notes to keep me warm in winter. The geographers put names on the coast but the ones not on the charts are even more interesting. These are the names fishermen have for spots. After you leave Port Hardy and head down the strait you pass many fishing spots with great names. The top end of Malcolm Island is called "the North Shore" and, by some people, "Korea." In Double Bay one spot is called "Finn Rock" because that is where the Finns from Sointula fished. Another spot is called the "Blue Line" from an old fishing boundary. "Dong Chong Point" is named after a merchant from Alert Bay who had a logging camp there in the 1930s. In Baronet Pass is the famous (or infamous) "Bank," so named because it opens only for a few hours, meaning the tide there is something else. A little farther down is the "Pig Ranch," and so it goes.

The Vancouver Island shore has the most setting spots for seiners, so it also has the most names. Some names fishermen use follow the charts. The "Blinkhorn" setting spot obviously refers to the light on Blinkhorn Island. All fishermen know this place as Blinkhorn Island, not Blinkhorn Peninsula as it says on the charts—but not one in ten knows about Mr. Blinkhorn and the Hawaiians he imported into BC as loggers. Or even anything about his daughter, who apparently was a remarkable woman. But fishermen know about "Hilly's Camp" just below Blinkhorn. This setting spot was where a cougar attacked a female logging camp cook and the boss logger attacked the cougar. The joke was that it was not a fair fight because the logger, in his rush to save the woman, did not have time to put in his teeth. Below that is Izumi Rock which earned its name the hard way, by one of the *Izumis* colliding with it. These are familiar names to strait fishermen.

The reasons for the names of some setting spots are clear, such as "Fine Beach" and "Glorious Bay." Others are not, such as "Pineapple." In the end, to make money it is not enough to know the names, you also have to know how to fish these spots. In some cases the penalty is small—no fish and wasted effort. In other cases, the net is snagged and ripped up, or even lost. However, after more than sixty years of salmon fishing in the strait most spots are well defined. If you are new to a spot, one way to learn is to watch others. If there is no one there (unlikely), the marks on the trees and leftover pieces of rope straps are indicators that someone has tied up the end of their net there, so it must be a setting spot. Nowadays, sounders and good charts are all you need. But the first skippers to set in a new spot must have had their hearts in their mouths as they explored new territory. What they saw and felt sometimes gave the name to the place, just like the early explorers of the coast.

My notes and journals are full of how to set at all these spots, but fishing time is so limited these days that not much of it is of any use any more. You have to set where you can, not where you want. The only time left now to reflect upon the names of setting spots is in the winter. And the names lead to more reflection and musings. What about the woman attacked by the cougar at Hilly's Camp, or Miss Blinkhorn? What ever happened to them? What about me and fishing? Will I see these spots again? Will they keep their names or will new people give them new ones? And what about T. Masterman Hardy? When it came time for him to die, who came forward to give *him* a kiss?

WINTER

INTRODUCTION

On the Beach

Winter in the fishing industry comes when the net is stored and all the gear is put away in the locker. Soon after comes the magic moment called the "settlement cheque." This is the final reconciliation of what you made from fishing, after your expenses and your cash advances. In little black marks on a piece of paper a season of your economic life is neatly tabulated. What the neat figures don't show are the happiness and heartache that went into it all. The settlement cheque and attached statement tell the story from a bookkeeper's perspective. Amount of fish landed, price received, grub and fuel expenses, cash advances, Unemployment Insurance and Worker's

Compensation deductions are the items that catalogue your financial well-being.

There are no deductions for income tax, though. Under a curious aspect of the law, fishermen are not employees, so there are no automatic deductions of this tax. This of course leads to much trouble. I am able to profit from this trouble as sooner or later—usually later—a fisherman will show up with a shoebox full of paper and ask me to do his tax because he has been receiving these letters from Revenue Canada. As I fire up my computer, sharpen my 2H pencils and start sorting out the paper in the shoebox I feel that I am truly "on the beach," which is a fisherman's term for when he is not fishing, and that for me winter has officially started.

To the outsider, there are two great mysteries to the fishing life: what fishermen do when they are fishing, and what they do with all the time they have when they are not fishing.

How do we spend our time on the beach? That is largely determined by the bottom line of our settlement. In a good year, many fishermen's thoughts turn to travel. The preferred destinations are the usual: Hawaii ("Big Island or Maui?" is the standard query), Mexico or Palm Springs. In bad years—of which there have been too many lately—we stay at home and take in simple pleasures. These simple pleasures vary with the individual but basically we all have to do a lot of what is called "catching up." Being away from home for several months every year means that many aspects of ordinary life get neglected and must be addressed while you're "on the beach." Being by nature handy, many fishermen take on home improvement chores such as fixing porches, painting and remodelling. Vehicle maintenance is another time user. For many fishermen, being "on the beach" means spending as much time as possible in their pickup trucks. For moving gear around, the pickup is indispensable and, better yet, tax deductible. Convenience and tax deductions are a happy marriage for us pickup lovers, which brings us to another part of the catching up process: family. Broken marriages are no surprise. The life we lead is full of uncertainty and the popular wisdom is that families need stability, although one wife told me the problem in a much simpler way: you get used to your mate not being home for long periods of time, and suddenly he's at home every minute of the day. Understandably, that can be very trying for all concerned.

It may be different depending on where you live. In the fishing towns and villages along the coast there are a lot of people in the same fix every winter, so there is some sort of common understanding about what is going on. It is time to socialize, but it is also time to get things ready for the next season. Everything moves at a slow pace and time is not a precious commodity. The government dock is deserted because all the fishing activity has moved to the local coffee shop.

Village life was (and maybe still is) a simpler thing than city life. I imagine in the old days—whenever that was—"on the beach" meant it was time to smoke dog salmon, cut wood and can preserves. We may do less of those things now, but we get ready for the winter in our own way, so the process is alive and well. Despite our trucks, cable TV, cellular phones and shopping malls, we are doing what fishermen have always done.

Old-time sailors called the winter a time for "make and mend." A time to make new things and mend the old. Nowadays there is not much to make as everything in the high-tech world of fishing is steel and plastic. Once nets are made up, they don't need much care. The same applies to the newer boats. Whereas a wooden boat needs extensive annual maintenance, the new aluminum boats don't even need paint. So all those traditional activities are getting scarcer.

This leaves more time for other things. Some fishermen I know are secret artists, others take up golf and others drift into fisheries that go on during the winter, such as the trawlers. Many skills may be no longer necessary, but new ones have arisen. Net mending and rope splicing are still around but now, knowing how to operate a sonar to catch schools of spawning herring is just as important. And just as in "the old days," these skills are learned on the job. No one I know goes to school to learn anything about fishing. This leaves time for much else.

When I first started fishing, I was impatient for winter to be over, but now I don't mind it at all. I have an awful lot to do. I am a kind of pack rat, a compulsive note-taker, and a collector of scraps of paper, all to do with the fishing industry. I have kept all my old settlement statements and income tax forms, because to me each of them tells its own story. I also have all my log books, in which I used to write a daily commentary. Onto each page I pasted a cutout from the tide book of the tides for that day. Now it helps me recapture how it was on that particular day on the grounds—what we caught and what I was thinking about. I also kept a personal diary where I wrote down the things that I was really concerned about and also, sadly, some rather bad poetry. I tried my hand at that, but gave it up once I realized that the most boring people I knew were poets.

In the box where I keep all this are also newspaper clippings that I've saved and photographs that I took. All this was done partly in an attempt to categorize the various fisheries and record the details of my experiences with them. The halibut file is very slim because I only tried that once. Beam trawling for shrimp has good pictures but the settlement statements make clear why I left that fishery. My favourites are my old notes on salmon seining with, if I do say so myself, rather nice little diagrams of how to set in various places on the coast. My notes on salmon gillnetting are sparse, because most of that fishery takes place at night, so I could never see well enough to write down any information. The biggest and fattest file is for salmon trolling. It is my favourite and it is well thumbed.

I still take notes, and winter evenings after the hockey game I turn on my computer and write them all up. My new project is to get one of those computer scanners and computerize my photographs and diagrams.

It is all interesting to me, but I am enough of a realist to see that there are a couple of flaws in what I am doing. Very few people have written down what it is like to be a fisherman. Why this is I don't know, but it makes me think that maybe I am kidding myself that anyone is interested in fishing history and the fishing life. The other problem is my perspective. In the early days I had a certain status as a boat-owner with my husband. We had progressed to buying our own small seiner, so I felt that I could talk to my fellow skippers and boat-owners with some authority. That all went down the tubes with my divorce, so now I can only write from the viewpoint of a lowly crewman who has to struggle for a job every year.

Another irony is that I love salmon fishing, which has always been a part of my life, but most of my money comes from herring roe fishing. I depend on it, but somehow I can't get excited about it, so the file on herring roe stays very small. Consistency is not my forte, I guess.

If I were a realist I would recognize that each year my hold on the fisheries gets weaker and weaker and that I am spending more and more time on the beach. Maybe I do realize this, and writing and updating my scrapbooks, although it gives me pleasure, is also a kind of self-defence. The pleasure part of it is half nostalgia for "the good old days" and half a simple love of the knowledge that is there to be had if one takes the time. I can look at a diagram of a set at Addenbrooke Light and recall the crew, the tide, how much fish we caught and what I was thinking at the time. I also know that Addenbrooke Point was named by Captain Vancouver after his officers came through that little pass from Rivers Inlet, which they called Schooner Passage. One glance at my little sketch and I can capture my life there and that of many others going back a couple of hundred years. I am in good company, I know.

The self-defence part of it is that even if I can no longer give my best as a fisherman—and I have to face the fact that my chance to do that may have come and gone—at least with all my notes and records and writings I am making some kind of a contribution. Maybe we fishermen grow nothing and produce nothing, maybe we start the year with nothing and end it with nothing, but in the meantime we have stood up under the sun or in the rain and lived life as few people can or do. In my log books and files and photos, I can make a record of this.

It is late at night in my room in the city. I am a fisherman on the beach, with not much of a future if I read current conditions correctly. But all fishermen are on the beach for a time; the problem for most of us is that that time is getting longer.

This should sadden me but it doesn't. Fishermen are fishermen for a curious mix of reasons, difficult to fathom at times. I know why I happened into it and why I stay. A bit of time on the beach merely means I am getting my gear and my soul in shape for the new season. There are more fisheries to explore, more places to set our nets, more treasures about the history of the coast and its fisheries. This will require more notes, more typing, more research. Sometimes I think winter is not long enough. Life on the beach has its own rewards.

The Muggins Rule

Winter is kind of an empty time for us fishermen, so we seek diversions. We need them. No money, no fish and nothing to do. And sometimes when there is nothing to do we play crib.

Crib, or more correctly cribbage, is a card game that I guess every fisherman knows how to play. It requires only a standard pack of cards, a counting method which is usually a board with a bunch of holes punched in it, and little pegs to record your progress along the board until you reach "home." Along the way you pass the "skunk line" (the penalty is double) and you do not wish to "die in the stinkhole."

Crib is meant to be a friendly game played for diversion, which rewards skill and luck, but not always in equal proportions—which makes it fascinating, sort of like life.

Last winter I played a lot of crib. I had lots of time on my hands, as did the entire salmon and herring fleet. We—that is, the industry—have had the big flurry of meetings about what to do about the industry's problems and we

The Muggins Rule

played the usual game of waiting for news from the mysterious east—Ottawa—where it seems our future is constantly being decided. We are sort of like an outpost of imperial Rome waiting for the latest edict from a Caesar, except that where the Romans got bread and circuses, we have Kraft dinner and crib.

As I said, crib can be a friendly game and it usually is. There are a number of variations on the standard game which give it variety. You can play solitaire crib against a mythical opponent, or you can play slow crib where you try to get as few points as possible, or you can even play nineteen crib, which gives you nineteen points if you have a "bust hand." These games are friendly, or at least as friendly as the players make them.

However, there is a version of crib that is deadly. It involves something called the Muggins Rule, which gives rise to bad feelings and strong words. Uncle Ed and I have been playing crib using the Muggins Rule and life has been fierce, with no quarter asked for and none given.

In crib you progress across the board by counting your points from the cards in your hand and then moving your peg from hole to hole. Your opponent then takes his turn and does the same thing. Counting your points in crib is complex, involving runs, combinations of fifteen and flushes, and things to do with jacks (called "his Nobs"). The potential for miscounting your hand is ever-present. The Muggins Rule allows you to take points in your opponent's hand that they missed. What began as a friendly game now becomes a contest of intimidation and humiliation. You dare not miscount your hand; if you do, the penalty is double and the lost points go to your opponent.

Usually only seasoned players invoke the Muggins Rule and surely Ed and I qualify. I must have played a million games in my life and Ed double that. Sometimes we play four-handed but usually it is just the two of us fighting it out over the card table. Sometimes we pause for coffee or tea (since he's gotten married Ed has switched from instant coffee to tea), and we may gossip or swap a few lies.

Ed's views are always contrary to the wisdom of the day, so his contributions are interesting and not always wrong either. For example, I was talking about the roundtable discussions the industry has been having about its latest problems. I asked Ed what he thought about them. He called the roundtable a game of hokey pokey, which is a kid's singing game. Everyone gets in a circle, singing in unison and putting an arm or a leg into the centre, singing "put your little leg in" (or arm or hand or hip, and so on) and then taking it out singing "take your little leg out." The refrain is "do the hokey pokey, that's what it's all about."

That is Ed's honest opinion of roundtable discussions. I guess what he means is that it's all a game, as each part of the industry puts in a proposal and then withdraws it. At least that's what I think it means. Ed makes his statements short and to the point and you have to draw your own conclusions sometimes.

This got me thinking about the fishing industry in general—always a dangerous exercise, according to Ed. "It will fry your brain," he warns. Nonetheless, I have been around long enough to know something about the subject and to have opinions.

Thinking about the industry and why it is like it is, I have concluded that a larger Muggins Rule is at work in it. The Muggins Rule says that if you don't take your points, your opponents will. So it is in fishing—if you don't catch the fish, someone else will. As a result, fear and greed become the major motivators in the industry. This makes everyone very single-minded and very cranky to boot.

The Muggins Rule

I must admit I am tired out from all the different opinions about what to do in the industry. So tired that I even went to church over Christmas, as Ed's wife is big in her church and I don't mind putting in an appearance for her sake. I had a good laugh during the service. The minister intoned the familiar words "and the meek shall inherit the earth." The old guy behind me muttered under his breath, "Yes, but they won't keep it long." So even in church, the Muggins Rule applies.

Dry Rot

Right now, somewhere on the BC coast, a man is in a dark damp spot in a fishboat with a light and a knife, poking at some wood. It is usually in the stern of the boat, in the lazaret or in the fish hold, that he performs this strange practice. After a bit he emerges and if he says "No dry rot," everyone is relieved, because the alternative is too horrible to contemplate. "Dry rot" means your boat is rotting away under you and there is nothing you can do about it. I have seen the inspection ritual many times and have done it myself. To tell the truth I had no idea what it was I was doing. Now I know different.

Last week I went to see some hiking pals of mine, who were busy putting a new roof on a house. Winter is no time to do it but they were forced to. It came about this way: the old cedar shake roof had moss on it so the owner got a power washer to wash off the moss. In doing so he washed off his roof, because the shakes had rotted away. They looked fine on the outside but were rotten through and through. So a task

Dry Rot

force was assembled to put the roof on. One of the workers told me more than I wanted to know about the fungus that causes dry rot. The stuff that rots out cedar shakes on roofs is more or less the same growth that you have to look out for on boats. It's kind of interesting if you like that sort of knowledge.

The fungus is called *Merulius lacrymans*. If you know your Latin you know that *lacrymans* means "to weep." So there is a kind of irony here. The fungus got its name because it "weeps" a drop of water as part of the rotting process. If you have dry rot, you will weep also.

But dry rot is a bit of a misnomer, because the fungus needs moisture to get going. If the wood has less than 20 percent moisture you're okay. Oak has always been a favoured material for boats because it does not absorb much moisture. Different woods respond differently to moisture and I guess it was through trial and error that the best materials came to be known.

All this got me thinking about dry rot in other areas— such as in life. Dry rot works from the inside and while everything looks fine on the surface, things are pretty bad underneath. That's what happened to the guy's roof and why the pressure washer just blew it all away. The same thing can happen to people. My own experience with dry rot was my relationship with my former husband. Everyone thought everything was fine because on the surface all looked well. A little bit of pressure and poof—it's all over.

I have been thinking that something like that has happened in the political world. The dry rot has set in. While everything is fine on the surface there will soon be some weeping. This kind of thinking was kind of morbid, so I was pleased when my hiking friends invited me to a celebration in honour of their new roof. I took along a six-pack and some smoked sockeye and was ready to party.

Fishermen's parties are more or less the same. Everyone talks fish and gossips about those who are not there. Character assassination, slander and outright lying are all acceptable. Hikers' parties are different. Many don't drink, most are physically fit and some kind of health nut is always lurking in the background. But I find the hikers to be interesting people, as if they have some kind of inner peace that comes from hiking.

Of course, at the party we got to talking about dry rot, mushrooms and other fungi, and I soon trotted out my theory that our political institutions and many of our social organizations suffer from dry rot, too. Well, this was too morbid for the hikers, who are by nature a distressingly cheerful bunch. I pointed out how imperilled our fisheries were, what with pollution, too many people wanting the fish, low prices and so on. The hikers would have none of it. They thought I should face the facts: BC is paradise. The fish will always come as long as we don't screw up the spawning grounds, you can always make some sort of living if you are willing to go up the coast and work, and so on.

So it was a standoff. One of the guys noted that maybe dry rot had gotten into my brain. What he meant was that this could explain why whenever I saw a problem I started sniffing dry rot everywhere.

"Focus on the doughnut, not on the hole," he said.

Somehow that cheered me up. Thinking it all over, I came to the conclusion that at least in a fishboat you know when you have dry rot. In life, it's harder to tell.

The Night Pete Shot Santa Claus

Every year there are certain obvious things that I do at Christmas time. But some years, like most fishermen, I have very little money because of a bad salmon season, and no prospects of seeing any until the herring season starts in March. With luck, I'll have enough UI stamps to qualify for pogey, which is what some call Unemployment Insurance. Times are tough, but at Christmas you have to put on a brave face and get on with having a good time.

One thing Uncle Ed and I did last Christmas was go to Ladner. Ed has some pals over there and we were going to get some old moose meat. These guys go hunting every year for a moose, and between moose hunts it's all they talk about. While they are fishing they all talk of moose hunting so I guess when they are moose hunting they talk about fishing.

Anyhow, they had some of last year's moose meat in their freezer and Ed and I were going to get it. I am not big on moose meat, but I went along as I had a hidden plan. These guys also had smoked Alaska black cod (which the

Japanese call sablefish and pay the big bucks for). I saw some in Safeway and it was about 15 bucks a pound. I really like it and usually like to eat it at Christmas. So I went with Ed for the moose meat, hoping that somehow or other the talk would get around to black cod and they would have to offer us some, and it would be a nice Christmas treat for me.

Now, to get from Steveston to Ladner you have to cross the Fraser River, which you do by driving through the Massey Tunnel. This always gives me the feeling of going into a foreign country, and maybe that's what Ladner is. Ed's pals call themselves "river rats" because they live along River Road next to the dyke. We got to the house and sure enough there were some "river rats" hanging around. We went out to the garage and opened the freezer and there was the moose meat all wrapped and dated. Next to it were the objects of my desire: packets of black cod that I could just tell were of the finest quality, even though they were neatly wrapped in plastic. I murmured that it sure looked nice.

"It's good stuff," our host agreed. "I'd offer you some, but it's Pete's fish."

At the mention of Pete's name everyone broke out into smiles and sort of chuckled. I didn't say anything but I could tell that, good or bad, there was something special about Pete.

We got the moose meat but no black cod. On the way back to Steveston I asked Ed about Pete. Well, this was a subject Ed could warm to. Ed hates Christmas and with obvious relish he told me about the night Pete shot Santa Claus.

Several years ago, when Pete's two kids were about five and seven years old, he came home early one evening to look after them so his wife could go out Christmas shopping. Pete had been into the red wine and was in that strange state between being very drunk and the hangover machinery kicking in. The kids were excited about Christmas as kids that age always are. The little girl was more than excited, as her birthday was close to Christmas and she started telling Pete what she wanted for Christmas and also what she wanted for her birthday. The list was quite long and of course the little boy added his list to hers. This somehow or other rubbed Pete the wrong way and he decided to take drastic action. Ed started laughing and hooting and snorting as he got into this part of the story and in his inimitable style acted out the part of Pete.

Pete went to the phone and hollered into it, "Operator, get me the North Pole." After a pause he said, "Put me through to Santa Claus."

The children watched in disbelief. Then Pete said, "Hello, Santa? There is a problem in Ladner here with my kids. Meet me in the backyard in five minutes."

He slammed down the phone, gave the kids a fierce look and went upstairs. He soon reappeared carrying his shotgun. The children's eyes bulged out in terror and amazement. Unknown to them, Pete also had a sea lion bomb in his hand. These are very large firecrackers that fishermen use to scare sea lions away from their nets. Once outside the house, Pete lit the sea lion bomb and it went off with a large bang. Pete re-entered the house and announced to the children, "I've just shot Santa Claus! Christmas is cancelled."

The impact on the kids was terrible. They started crying and shouting at Pete that he was a killer and an all-around bad person. Of course, Pete's wife came home as the kids were crying and I imagine one of the all-time great rows must have broken out. What a scene: crying kids, angry wife, unrepentant Pete. I was horrified but Ed seemed to think that this was one of the greatest stunts of all time.

Heading home through the Massey Tunnel, with no money and no black cod, I wondered if maybe Pete did shoot

Santa Claus. I hoped to hear that he came to a bad end, with his wife and kids leaving him, but that is not how they do things in Ladner it seems. Pete has not changed, his wife is still with him, and the kids are older and wiser—at least, they certainly no longer believe in Santa Claus.

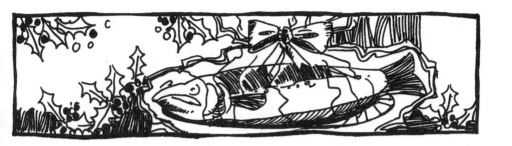

How to Travel with a Salmon at Christmas

Christmas means many things to many people. To just about everyone it means turkey and gift-giving. Among my friends and relatives, Christmas sometimes means giving a salmon. The principle behind this is obvious: those of us who have salmon give it to those who don't. On the face of it a simple gesture, but it really is more complicated than that.

If you ask me, we are potlatching. The potlatch is a West Coast Native Indian custom whereby the wealthy gain prestige not by accumulating wealth, but by giving it away. So if you have lots of salmon you gain prestige by giving it away, and you also have a happier community around you, because now everyone has salmon.

But it doesn't end there. Each receiver has now been put in the position of appearing less important than the giver. So all the receivers, in order to show that they too have wealth and prestige, give gifts back to the giver. And on it goes. Sometimes I think that the potlatch is so much like Christmas that it is hard to tell which tradition is which.

Anyhow, I am either into Christmas gift-giving or potlatching, as Christmas usually finds me going over to the Island with some salmon. Travelling with a salmon at Christmas is a chore only for the truly determined, or maybe I should say demented. It is fraught with peril.

The salmon is caught in the summer, cleaned and frozen. Now to freeze a salmon so that it will be nice when it is defrosted requires some care. You need a good freezer and a bit of time. You dip the salmon in water and freeze it. Then you dip it in water again and freeze it again. You repeat this process until you are satisfied that the salmon is completely airtight, since contact with air is what causes freezer burn.

Now your salmon can be kept until gift-giving time at Christmas. This is not the tough part. The tough part is delivering it.

In the summer, delivery is no problem. After years of experience, you've had enough of being suckered into cleaning a salmon for a friend, then delivering it to the friend's house, then cleaning out the friend's fridge to find space for it. You don't do that any more. You now force the receiver to come to the boat and pick up the salmon. Those who truly want it will come and consider themselves lucky. But in winter it is different.

You now have this frozen salmon which is about a foot and half long—a size predetermined by God, or perhaps by some failed evolutionary strategy, not to fit into any existing man-made container. Somehow or other you have to get a box big enough for it and then a new problem arises. As soon as you put in a bunch of ice cubes, you know they are going to start melting, so the box has to be waterproof. And you can't let them melt too much, so you need to insulate the box. The alternative to the melting-ice problem is to get some "dry ice," which is frozen carbon dioxide, and use that. But quick now—where do you get it? No one knows.

Assuming you have the salmon in a suitable box and some method of keeping it frozen, you are now ready to travel. Except that now that you've taken all the right precautions, the damn box is as big as an old-time steamer trunk. It won't fit anywhere in the car except the trunk, so you have to travel light in order to make space for the salmon.

Having sorted all that out, you are ready to head for the ferry to the island or wherever else you are going. If it is going to take some time to get there, you are on the lookout for preservative measures. Try checking into a motel and asking them for a freezer or cooler big enough for you to store your salmon. They don't have them. Butcher shops have them, but they are not partial to assisting passing strangers with their frozen salmon. In desperation I once thought of saying I had an organ transplant in the box, but I would be amazed if anyone ever believed me.

Travelling at Christmas is already a strain and travelling with a salmon makes it even more tense. As a giver of salmon, your objective is speed and you are concerned, even obsessed, with time and temperature. Your companions may want to enjoy the trip and maybe even be planning to take their time. You must not allow this or any other attempts on their part to delay. Once you have the salmon in your possession you are chained to it and it dominates almost your every waking moment. Your mind continually races with frantic thoughts. The weather is too warm and the ice will melt, the ferry is too slow, or you didn't get enough ice to begin with. With all this stress, soon you start thinking evil thoughts about those who are to receive the gift of a salmon at Christmas.

These thoughts are definitely un-Christian. Probably the

people who are going to receive this salmon already have some and won't appreciate yours. Or if they are a large family, as soon as you give them one fish, they will think you should have given them two and wonder why you didn't. Maybe they don't know how to cook it properly and will ruin it.

Now you start cursing yourself for being such a fool as to take the time to freeze the salmon correctly, keep it for several months, pack it carefully and guard it with your life on its final journey. For what? For a bunch of jerks who maybe are your relatives, but who have never appreciated the fishing life.

But somehow it always happens that you arrive on time and find that the salmon is still perfectly frozen, ready to be admired by one and all and then slowly defrosted in the fridge (the correct way). The world starts looking brighter as you relax over an eggnog or a rum punch, but you look closely into the faces around you. You cannot detect that they are anything but pleased to see you and to have been given this salmon, yes a lovely salmon, caught by you just for them. In fact if the shoe was on the other foot, what could they give you that would ever be as good as this?

As you sit back and relax you realize the answer. Nothing. There is nothing that is as good as this salmon. You have just found out the true meaning of wealth. Wealth and prestige. So it was worth it, after all.

The Year of the Two Santa Clauses

I don't get all that excited about Christmas, but I don't mind it. To me, it's a nice break from the routine. Uncle Ed, on the other hand, actually hates Christmas. He doesn't shout "humbug" or anything like that, he just doesn't say anything. That is a sure sign with Ed that he doesn't like something.

I never knew why he hated Christmas until I met a guy who grew up with Ed. He told me the story of the Year of the Two Santa Clauses.

You have to remember that Ed grew up during the forties and fifties and life was much simpler then, especially if you lived in a little town on Vancouver Island. Salmon fishing ended in late November, after the Nimpkish and Satellite dog salmon runs. Wood was cut, winter came and the southeasters blew. In those days, the Christmas concert at the church was a big deal because there was no TV. It was *the* winter celebration in the village—a tradition that everyone participated in.

One of the high points of the concert was the arrival of

Santa Claus, who gave out gifts to all the children. The concert was put on by the school, but it took place at the local church hall, with the support of the one other church in town. Each church had a Santa suit, which could be borrowed by the school or whoever else needed them during the Christmas season.

Many aspired to wear the revered red outfit. Santas-in-waiting trained by appearing at parties and in the elementary school and put in years of service to work their way up. The senior position was to be Santa at the annual Christmas concert.

The church that Ed's family went to had a resident Santa, Al Johnson, whose position was coveted by one Bill Williams. Because of their rivalry these two men disliked each other intensely, even though they belonged to the same church.

On the fateful year in question, Uncle Ed and many other little darlings like him were gathered with their parents in the church hall for the annual concert, ready for the festivities to begin. All the kids eagerly awaited Santa and his big bag full of gifts.

But trouble was brewing. As he did every year, Al Johnson had the church's Santa suit and was getting prepared for his annual grand performance. Tragically for him and others, he stopped in at the beer parlour to buy some cigarettes. Someone saw him, and a few phone calls later, the news reached Bill Williams that the church's Santa was drunk, and that this year's concert was going to be ruined.

Those of you who have lived in a small town know how this can happen in a moment. Bill felt greatness being thrust upon him and there was only one thing he could do: borrow the other church's red suit and rescue the honour of the church and all Santas, past, present and future—and at the same time, with Al Johnson's reputation in ruins, supplant his rival as the official Santa for the years to come.

The Year of the Two Santa Clauses

Of course, all this was news to the present Santa, Al Johnson, who had made the briefest of stops at the beer parlour and was proceeding in his measured way to get ready and make his grand entrance. To do this, he had to take his suit up a ladder at the back of the church hall, climb through a window, dress in darkness, wait to be announced by the master of ceremonies and finally make his grand entrance onstage. As far as Al was concerned, everything was going according to plan. He was unaware of the parallel efforts of Bill Williams until, moments before the curtains opened, he heard a sound in the darkness behind him.

All who were present on that day can recall exactly what they were wearing, what they saw, who did what and so on. Except that everyone's version is different. To this day, there are those who maintain that Al was drunk, others that he was not. Equally, some maintain that Bill only showed up to save the church concert, not to grab the Santa job for himself. Whatever the truth, some facts are not in dispute. That the children, Ed among them, were of an age that still believed in Santa is true. That the church hall was full is also true.

This, then, was the full horror of that time so long ago. When Santa's arrival was announced to the children and the curtains parted for the traditional "Ho ho ho," a much different sight presented itself to the eager audience. What was revealed was not one but *two* Santas, and they were trading punches and cursing as they wrestled about on the stage. It was clear that they both had got dressed in darkness and appeared together at the climactic moment.

Of course, it was all sorted out. I am told that Bill Williams never got to be Santa, and that Al Johnson never took another drink until his dying day. But the Christmas concerts in Ed's village were never the same again. As for Ed and the other children in the audience on that fateful day—

well, they never believed in Santa Claus from that moment on. A generation had lost its innocence.

I wasn't there myself, and it all happened a long time ago, but I guess the Year of the Two Santa Clauses was enough Christmas to last Ed a lifetime.

A Fisherman's Frolic

When the festive season arrives—by this I mean Christmas—all fishermen who have been good little boys and girls expect to receive nice things. As for the rest of us, we'll take what we get, make no complaints, and close our eyes until it's over.

Like Scrooge I've been thinking of Christmases past. It was actually Uncle Ed who got me going on this subject, and got me thinking about the differences between how we do it all now and how it was before.

We were in Ladner and drove by that little fishermen's hall just off River Road. I idly wondered if there would be a New Year's or other kind of party there this year. Ed noted that there was nothing like the "fisherman's frolics" that they used to have in the good old days. Because I had nothing better to do, I pressed him for details. After complaining that no one knew how to have fun any more, he gave me very specific details about the typical fisherman's frolic that they used to have upcoast in the fifties and sixties.

These were well-organized soirees which were put on around Christmas or New Year's or even a bit earlier to celebrate the end of fishing season. Usually the fishermen's union or the local co-op or some other organization sponsored these dances, which were just called "the annual dance," or had names like the "Welcome Home Party" or "Fishermen's Homecoming." On the island when Ed was a young man it was always called the "Fisherman's Frolic." In general, everything went like clockwork and everyone in the community showed up at them.

It was the composition of the bands at these affairs that intrigued me. A piano, a set of drums, a saxophone or two, an accordion and occasionally a violin. No guitars, electric or otherwise, and the piano player was always a woman.

The music too was special—played in sets of three tunes apiece, because the convention was that you asked someone to dance for a "set." Thus, three waltzes, three polkas (various speeds), three schottisches and God help me, even French minuets. A novelty number would be a spot dance, a broom dance, a "Charlie McCarthy" (a goofy dance where you sit on your partner's knee at those words) and so on. People knew how to dance in those days. And everyone danced. Women danced together, kids danced with their grandparents, fathers with their daughters and so on. Some of the favourite tunes were "In the Mood" (usually an excuse for some slightly scandalous jitter-bugging), "Goodnight Irene" and always, at the end of the dance, "Hawaiian Farewell" ("... sooon you'll return ..."). The way Ed described it, it sure sounded like fun.

I asked a few other old-timers and they more or less confirmed Ed's description of these fishermen's dances. Then one of Ed's old cronies asked, "Did Ed tell you about the problem he had with the logger's wife and the trouble with the Mountie?" My ears perked up. I knew I was going to get a long story but it sure sounded juicy.

It was a long time ago, when Ed was in what he now calls his "wild stage." In the village he lived in, at the end of the salmon season the fishermen had lots of time on their hands. Meanwhile, the loggers kept to their usual routine: in camp for ten days and at home for four.

Apparently Ed took a shine to a logger's wife, and started dropping by to visit when her husband was in camp. You can't have a secret life in an island village, and soon there was gossip going around about the two of them. So much so that even the logger himself started to hear rumours.

Everything came to a head at the dance that the union put on, called "the Annual Fisherman's Frolic." The logger was there with his wife, but Ed was dancing with the older women and otherwise acting more than usually angelic. But his comeuppance was about to arrive.

Even though the dance had a liquor licence, some of the traditionalists still went outside the hall to have a drink and unwind. Ed was out there with his pals when suddenly he was confronted by the irate husband of the woman in question. The logger clearly wanted to fight. "Come on Ed, put 'em up. You been fooling around with my wife!" he challenged.

Ed tried to maintain his air of quiet dignity. "No I haven't," he replied, and it might have all ended there but Ed, being Ed, didn't like being challenged so he added: "I wasn't fooling around, I was *serious*."

Of course that tore it. The logger lunged at Ed and they started in on the traditional melee called "the fight outside the dance hall." However, the logger and Ed were quickly separated and Ed was hustled away into the hall.

Although he had made his point and got in a few good licks, the logger was not satisfied. Within moments he vanished into the night, and although no one knows how the rumour started, soon it spread through the crowd that the logger had gone home to get a gun to shoot Ed.

In the midst of all this excitement, someone had the presence of mind to try and find the local RCMP constable. This particular Mountie had planned on spending a quiet New Year's Eve with his girlfriend, and you can imagine he was not very pleased at having to come out and attend to the problems caused by Ed.

The Mountie came, asked a few questions and went to Ed. He gave Ed a choice: he could go to jail for a week, or he could go home. But if he went home, he had to *pretend* that the Mountie had arrested him and he had to act as if he was in jail. If he didn't agree, he would really go to jail.

An arrangement like this might not be by the book, but the Mountie probably figured it was the best way of keeping the peace, and it had the added bonus of allowing him to spend the rest of the night with his girlfriend, instead of with Ed.

Ed agreed. He went home and stayed there for a week. And so the story ended. The next time I saw Ed, I was dying to press him for more details, but I didn't. Ed always claims that he only remembers the good times.

The Year of the Pig

Since my world revolves around fishing, I like to think that I live in Steveston. But I can't find anyone who knows exactly where Steveston ends and Richmond begins. I really live in Richmond, I guess, and that means part of my world is Chinese.

Over the last few years, a big influx of people from Hong Kong has changed Richmond. We have a big shopping centre here that specializes in the Chinese market, and once when I dropped by there I noticed in front of it a large, happy looking pig about five feet long. This was in 1995 which was, under the Chinese calendar, the Year of The Pig. The Year of the Pig is supposed to be one of prosperity, because if you were a Chinese peasant the ownership of a pig meant you were well off.

Not everyone believes in these symbols but this one I like. First, I want prosperity—who doesn't?—but more than that, this sort of thing helps give a flavour to everyday life. Knowing that it is the Year of the Pig and that things will be prosperous kind of puts a little bounce in my step.

So I was in a chipper mood as I did my rounds. I go to the bank, to the supermarket, to the mall, check out the waterfront, maybe a drive by Paramount net loft or the "pond" where the gillnetters are, and drift through Steveston looking for people and gossip. It's a nice little route that is always the same but always different.

In the Net Shed Cafe in Steveston I ran into the Scientific Fisherman and his pals. Scientific is always good for a long, involved explanation of things he knows nothing about. His preamble always includes the necessity of being scientific, and today I was hoping he would be in a scientific mood. But he wasn't. He was listening to one of his pals talk about a cook called Little Joey. You don't have to be a rocket scientist to figure that this guy was small, he cooked on a seine boat and his name was Joe. Now Little Joey was a good cook, as cooks go, but his great claim to fame was that the crews always ate all of his meals with great gusto. His method was quite simple: skip the noon meal and force the crew to subsist on mug-ups until supper which was always served late. By then I am sure they would eat a donkey stuffed with firecrackers.

Trying to get in a word about the Year of the Pig and prosperity, I asked if Little Joey ever made enough money to open a restaurant. Well, his problem was that his timing was always wrong. If he decided not to fish halibut that would be the year that landings and prices were up, or if he left Boat A for Boat B, sure as sin Boat A would have a big season. He never could get it right.

But if financial rewards eluded Little Joey in fishing, he was lucky in another area, that of love. Perhaps a little too lucky, as at any one time several women would swear that

The Year of the Pig

they were engaged to Little Joey and that marriage was imminent. Joey was always able to delay on the grounds that he had no money and that as soon as he got a big score on the salmon or the halibut, wedding bells would ring.

What happened next, everyone swears is true—and if it isn't, it should be. The boat Joey was on had a couple of big weeks on the salmon. Joey's latest fiancée found out about this and put it to Joey: marriage or else. It seems incredible but Joey tried a goofy thing. To stall and buy some peace he convinced his loved one that the skipper could marry them on the boat and it would be legal. I think drink was involved here. Anyhow, at low tide the skipper of the boat married Joey and his bride.

I would like to report that it all turned out okay, but these things never do. For a while Joey seemed happy and a changed man. The boat caught fish, the wife got some of the money and Joey was decidedly more sober and respectable. Probably his cooking even improved. Joey's first mistake was that he got to fooling around with other women. His second mistake was that he started to let it quietly get around that the marriage was not a true one. He was hoping that once word got out, one thing would follow another and finally his new bride would accept the inevitable and move out.

You can imagine that intelligence was not high among all the parties involved here, but Joey was not the only one who could act with shrewdness, even with cunning, in order to get his own way. Joey's wife fought back. First, on the financial front she got Joey to sign a paper that allowed her to make a "drag" from the fishing company against his final fishing settlement cheque. So she made sure that as soon as Joey's boat made money, she got it. Second, she was somehow related to a Mountie, so one day a cruiser pulled up and a uniformed RCMP officer knocked on the door to have a quiet talk with Joey. In essence, he told Joey the marriage was legal and that he had better start acting like a married man or he, the Mountie, would take every opportunity to put Joey behind bars.

Now all this was long ago and I am sure the story has been amended in the telling and retelling but everyone swears that it is true. So I asked what became of Joey and his bride. Well, against his will or not, once his wife started managing his affairs Joey became prosperous and opened a little coffee shop. His wife worked in the front—and managed the till I am sure—and although Joey lived out his days the victim of his own scheme, for him every year was the Year of the Pig.

Roe, Roe, Roe Your Boat

As winter draws to a close, an unnatural quiet falls over Steveston. About the only excitement comes from the trade show at the exhibition grounds on East Hastings, where it's always fun to look over all the new equipment, even though it is usually just the old equipment with some new bells and whistles. The beer garden is full of salesmen, and herring seine skippers plotting against each other. This year, I got drawn into a discussion about the new designs in herring punts that seem to be coming out. The way I posed the question was: what is the most beautiful boat in the world? Beautiful to fishermen is the one that does the job and makes money, but everyone has different ideas about how to go about that. The result is that although herring punts sort of all look the same, they come in all different sizes with all kinds of variations in design.

Fishing vessel design is one of my interests. I even have the books published by the UN's Food and Agriculture Organization (FAO) on fishboats and a fat thick file full of drawings of old boats. I couldn't find too much in my file when I got back home from the trade show, so—maybe it was the beer or the excitement of the trade show—I suddenly decided to go to the library and see what I could find on the theory of fishboat design.

On the way to the library I stopped in at Uncle Ed's. He had a visitor—one of his old pals, George, a Norwegian fisherman who has retired and has always been an ardent communist. George is the only person I have ever known who not only has a copy of Karl Marx's *Das Kapital* but who has actually read it.

I knew better than to get into a discussion with George, but on my way out I offhandedly said to Ed that communism was now old stuff that didn't matter any more. Unpredictably, this made Ed mad as hell. "See you later," he snorted as I left.

The library had lots of weird stuff on fishing vessel design. I read everything I could find and made copies of the most interesting material although of course there was nothing about herring punts. The herring roe fishery has grown up so fast out of nowhere that fishermen just put together the most basic design they could think of and there has not been too much time for it to evolve.

You know a herring punt when you see it. It is made of aluminum, has a flat bottom, and certainly is not a sea-going vessel. It is designed for one job and one job only. Many punts have gotten modified for other jobs, such as clam digging, but essentially the herring punt is a single-purpose vessel.

In some of the new designs, owners have combined their salmon gillnet licences and their herring licences on the same vessel. The result is a new breed of gillnetter, a bow-picker built for speed. It can pass anything except a fuel oil dock.

Roe, Roe, Roe Your Boat

Now all these vessels were built for fishermen who I am sure have never read a book on vessel design or naval architecture. Why would they?—from experience they knew what worked and they had their boats built accordingly. I got to thinking about the skippers and owners of fishboats here in BC. From sounding them out, I knew that the theory of vessel design was not their favourite topic of discussion. Certainly they could pass judgment on various designs, but I felt that their analysis was far from rigorous. "That's a good boat" or "That's a bad boat" was about their limit, it seemed to me.

So, the gillnet herring punts that work in the roe fishery of BC have been designed pretty much without benefit of theory or principles other than basic common sense. Fishermen with just experience and a rough idea and welders skilled at working with aluminum have produced a design and type of vessel that for better or for worse is unique in the world. I was somewhat awestruck at my conclusion and immediately somewhat frustrated: I knew that there was no one I could tell this to who would be similarly impressed.

Before leaving the library I thought I might as well look up all the references to fishboats that I could, and I came across a dandy. One of my heroes of English literature—Sam Pepys, the diarist of the British Navy in the reign of Charles II—wrote about fishboat design in the North Sea herring fishery in 1669. Not only did Sam write well about everything, but this particular passage made me feel that he would be right at home in Steveston.

Sam was in charge of building new ships for the British Navy and he questioned everyone as to what was the best design. He was astounded that no one could give him a clear answer. Sam had asked why Dutch herring boats were different than British ones and he was not impressed with the answers he got from the Royal naval architects. So, like me, he did his own research. (By the way, Sam was motivated by more than professional curiosity. He wanted to be a partner in a royal monopoly for the herring fishery.) His research gave him his conclusion. He noted in his diary: "It is my observation that countries design vessels that suit their purposes and their coasts." In other words, fishing boats were designed by fishermen for a fishery they knew. That is why fishboats are all different.

Driving back from the library I decided to drop in on Uncle Ed and shoot the breeze a while, as I knew the old communist would be gone. Ed welcomed me in and said nothing about my previous remarks. While eating some leftover Christmas fruitcake I told him about my research and what I had found out.

I talked about what Sam Pepys said in 1669 about herring boats. All of a sudden Ed got that fierce look that I know so well.

"Dammit," he said, "you told my old pal George that communism was old stuff and it was no good and now you tell what this guy said three hundred years ago and it is supposed to be good stuff? Make up your damn mind."

I knew that I was pushing a lost cause. Ed and I owned a punt and he saw it as what it was, a way to make money. So what if I saw it as something else? Sam Pepys was right, but then so was Ed. As for George the old communist—what happens to dreams that die?

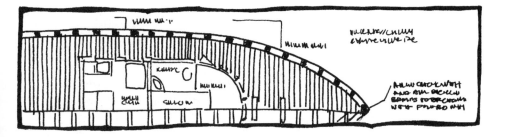

A Good Rule of Thumb

A good rule of thumb is … well, it's a good rule of thumb. I got to thinking about these handy little rules when the computer guy came over to load some new software into my computer. I'm getting ready for income tax time and decided to upgrade. Computer guys are worse than fishermen for using jargon and technical terms, but what he wanted me to do was buy some database software, and that got us talking about the need for information. I belong to the "less is more" school of thought and argued with him. We don't need more information, we need less of it. I noted that most databases are just loaded with useless junk. What we really need, I said, were good rules of thumb. A good rule of thumb minimizes the amount of information you need to make a decision.

Once I started thinking about rules of thumb I plowed through my notebooks to see if I had one or two buried away. I found I had to think about what a rule of thumb really is. First of all, I found out that the phrase comes from the practice of estimating a yard, which is a bit more than the distance from your nose to your thumb. Or at least from the nose to the thumb of whoever the guy was who invented the yard, but the point is, if you don't have a tape measure close to hand, you can use the one that's built into your body. Fishermen still stretch out their arms to get six feet which is a fathom, one of the commonest measurements in the industry (in fact until quite recently, charts still measured depths in fathoms).

One thing I had to do was to sort out the difference between mottoes, sayings, proverbs, ordinary rules and rules of thumb. The key difference is that a rule of thumb condenses all the information at hand into an immediate form that you can use to make a decision.

The best example of this is that old weather rule: "Red sky at night, sailor's delight; red sky at morning, sailor take warning." Mother Nature gives you the information and you have the rule. That's all you need to predict the weather. Fishermen really need these little rules, because there isn't any book on fishing that you can study and there certainly isn't any university that teaches this stuff. So fishermen have had to develop lots of little sayings and snippets of information about how to make decisions on when, where and how to fish.

I am the only person I know who writes down all the information I get about fishing in my notebooks. The problem is they are a mess. When I first started, everything was new and everything seemed important. Now I am at a loss to figure out what is important. One reason is that I know much more now and have a "feel" for how things should be done. Fishermen know lots, but they don't write it down, and in many cases they don't realize just how much they know.

This conclusion led me to a question: what is the one rule of thumb that all fishermen have? The answer was easy.

A Good Rule of Thumb

It wasn't in my notebooks, because it is so widely known to be true that even I hadn't bothered to write it down. The rule is: have your gear in the water at low-water slack. At first glance this does not seem to be a very profound thing. But I know it is. I asked around some of my pals who are in different fisheries and this rule applies across the board. Especially to salmon fishermen, but to a lesser extent to herring fishermen, to draggers and to the other smaller fisheries.

The reason why this rule is so important is simple. First of all, it tells you when to set your gear. The when determines the where and how. The rule gets you going in your search for fish and once you are working, things start to fall into place very quickly. Fish are information. If you gets lots, that's one thing (lucky you!) and if you get only a few, that's another thing.

There is also a common-sense reason for putting your gear in at low-water slack. The change of the tide to the flood gets fish going in a specific direction, and with luck sooner or later they will move by you. For example, generally the flood tide goes in the direction of the salmon spawning grounds, so the salmon will go with it. As a result, salmon fishermen are wired into the flood tide. The fishing strategy that they plot in their mind's eye is based mainly on the flood tide. Low water slack is the overture to the beginning of the flood. If your gear is in the water you are in business.

So you see why fishermen have these little rules and sayings. Everyone has their own versions and variations of them, such as "the third set at the merry-go-round is the best." While almost incomprehensible to outsiders, rules like these are the stuff of knowledge, experience, skill and ultimately success or failure. What separates fishermen from most people is that they are desperate for information at all times. They need to get information and interpret it in order to make their living. Good rules of thumb make all this easier.

Fishermen put their gear in the water at low-water slack because they know it works, and furthermore they have always done so. In Shakespeare's play *Julius Caesar*, Mark Antony says before the battle of Actium: "There is a tide in the affairs of men, which, taken at the flood, leads on to fortune …" Who would have thought that Shakespeare knew a fisherman's rule of thumb?

Oysters and St. Valentine

They say it was a brave man who ate the first oyster. But I know that once he ate it, he ate many more. Say what you will, they are delicious mollusks. I have eaten them smoked, fried, raw, stewed and even baked in a loaf of French bread.

I thought about all this as I went on the ferry from Horseshoe Bay to the Sechelt Peninsula, or as the locals like to call it, the Sunshine Coast. The sun wasn't shining but that didn't bother me. I knew I was in for a good time visiting my cousin Leroy on his oyster farm.

Okay, his name isn't really Leroy and he's not exactly my cousin. The name Leroy was of his own choosing. As a kid he found it meant "the king" in French, so he chose to switch over, finding Leroy much more regal than the name "George" that his mother had given him. He always felt himself to be a self-made individual, so why not a self-made name? As for Leroy being my cousin—well, he is actually more like a shirttail relative by marriage, but it's kind of a long story so I just call him my cousin.

Leroy is always interesting to be around. He has a personality beyond the ordinary, and somewhere he has picked up an education to match. But more than that, he has one of the sharpest wits I've ever met. Unfortunately, he started his career as a fisherman, and as a fisherman he was a bust. On a troller he tended to get seasick. On a gillnetter he exerted a peculiar magnetism that could attract the only drift log for miles around, and sea lions were his constant companions. As soon as he stepped aboard a seiner either all the fish in the neighbourhood disappeared, or the engine broke down, or both. Once, in the middle of a particularly bad season, he told me that he was going to take an oar out of his skiff and start walking, and that as soon as he got far enough inland that people would start asking him what that thing was he was packing, he would stop and settle down. Evidently that's how he got to the Sunshine Coast. It's not exactly inland, but at least oysters don't swim. Leroy felt he could cope with a species that stayed in one place.

After some years labouring in the oyster business, Leroy had bought his own oyster farm. I was going to set up his books for him, and he would explain the world of oysters to me.

I took the ferry over to see him on St. Valentine's Day—the day set aside for lovers. Oysters, of course, have a connection with that amorous saint, as they are considered an aphrodisiac. Everyone seems to know this, although no one, including companions I have shared oysters with, has ever offered me any actual proof.

Leroy met me at the ferry and we drove to his "farm." An oyster farm is really only a lease from the province to grow oysters on a piece of shoreline. As such it is mostly out of sight, at least at high tide. On the shore there will be a small shack to store tools and trays and stuff, and somewhere there is a skiff to go out to the floats from which are hung

Oysters and St. Valentine

oyster strings. In Leroy's case, he also had an old herring punt with an outboard.

At Leroy's place we set up his books. The finances of an oyster farm are not too complex. The capital equipment is the skiff and the outboard and the floats. The oyster spat (the "seed") is quite cheap. You can get into the oyster business with only a few bucks, but the trick is to make money at it. From what I could see, what with harvesting and so on, the business is highly labour intensive. For this reason oyster farms are usually family owned. Nobody gets rich, but then nobody starves either. Leroy's analysis was short and sweet: buy a farm and you buy a job for life.

Leroy brought me up to date on the salmon farmers, who are now mostly gone. In the 1980s a bunch of salmon farms started up, but for once the Sunshine Coast lived up to its name. Each summer, the sunshine brought with it a plankton bloom in the water that wiped out many of the salmon farmers. This did not affect the oysters, although the oyster farmers have had their share of problems with the red tide. Most of them survived because their operations were small and they had already found ways to diversify their incomes, so they could stand a season with little or no cash. If the red tide came, Leroy would find something else to do until it passed. Small can be beautiful, if you find beauty in a positive cash flow.

When we were done with the books we went to the local pub and I met some of Leroy's fellow oyster men. On the whole they were not like Leroy. They were older and somewhat slower. If you asked them how they were, they thought the question all the way through before replying. I guess the occupation attracts a certain type.

But the oyster farmers were not dull and because it was St. Valentine's Day the jokes about love and oysters were flying fast and furious. I got them going a bit by noting that the oysters they grew were not the true BC oyster, but imports from Japan. One guy countered with the comment that like most people, I knew next to nothing about oysters. He was right, but whenever anyone says this about any subject I simply dare them to enlighten me.

Did you know that an oyster will not die of a broken heart? The reason being that they can and do change their sex. They start as females, become males and then go back to being females again. This would be a boon to some people I know. If you're unsuccessful in one area, no problem—just go over to the other side. Apparently European oysters can change their sex indefinitely, but the big BC ones, which are Japanese imports, get only two chances. That's still one more than you and me. I think if I were an oyster, St. Valentine would be my patron saint.

The Wealthy Fisherman

Sad to say, most of us fishermen don't have much wealth. We do, however, have a life. The trick is to make the money you have get you the life you want. Poor people work for money, rich people make money work for them.

Now, I like to see people who make their money work for them. Not too many seem to be able to do it. The TV and the papers are full of people telling you how to become wealthy and giving you a whole bunch of rules. Some of it is useful I guess, but it's not for me. Fishermen have such different lives that most of the standard financial advice is of no use. For example, a regular savings plan—how can you have one when the amount of money, if any, you make each year is different and you never know in advance what it's going to be?

I lucked out last week. I met a wealthy fisherman. He was wealthy in two ways. One, he had lots of dough and two, he had a good life, or so it seemed to me. A big part of having a good life was that he no longer had that nagging anxiety about money that most of us fishermen live with on a daily basis. He knew he had a system that would give him what he wanted. I guess he had that rare thing, peace of mind. My next question was, what are the rules or habits that produce a wealthy fisherman?

I met the Wealthy Fisherman at Uncle Ed's place. I knew it was a big deal because Ed was sitting in the living room with him. Normally, Ed would be sitting in the kitchen playing solitaire on the table and watching the little TV in the corner. Sitting in the living room with the big TV off meant that heavy business was taking place. They both clammed up when I came in. Ed remarked that they were waiting for a phone call. I got the message: I was to make myself scarce when it came.

With an unknown quantity of time to kill, I chewed the fat with the Wealthy Fisherman, getting him to throw me a few softballs about the industry, knowing that sooner or later I would get some good stuff. To get wealthy you have to think about getting wealthy, and if you're thinking about it sooner or later you'll talk about it. And so he did.

It took a while, as he did not just come out and say "do this" or "do that." Rather, he made a few general observations about life, fishermen and boat prices and licences.

Generally, I gathered that the Wealthy Fisherman was what the finance people call a contrarian. He said that fishermen were like a bunch of sheep. If someone was buying herring licences, everyone else wanted to buy herring licences. Naturally, soon the price of a herring licence went up and people started getting stung.

The time to be selling, he said, was not when everyone else was buying. The top price was not the best strategy. You want to be known as the guy who sold too soon, rather than the guy who sold too late. Interesting, I thought. He's not talking about money so much as he's talking about human behaviour.

The Wealthy Fisherman

For example, after a big season everyone rushes out and buys licences or quotas. Wrong. Hang on for a year or two and buy the bargains if you want licences or boats. Or better yet, put your money away. The dream of most fishermen is to own an apartment block, preferably in Hawaii. But how the hell can a fisherman weigh property values, taxes, insurance and maintenance costs on an apartment block if he can't even strike a good deal for a licence in his own industry? A fool and his money are welcome everywhere. Avoid conventional thinking, and if you know your industry, you should be able to find value where no one else does.

The Wealthy Fisherman noted in passing that all the best fishermen come from non-fishing families. I wanted to argue that one, but instead I kept my mouth shut because I knew the point he was trying to make: original thinking is the key to success.

Another thing he said was, "Get next to the money." I pondered that one. Then he elaborated: if you can figure out how the money goes around in the fishery you can figure out how to profit from it. He had a reputation for quality fish, which he sold for a good price to cash buyers. He had figured out where the money was to be made.

But there was more. Fishermen need good bankers, bankers who have the money and will give it to you. But you have to make it easy for them. The Wealthy Fisherman tallied the number of guys who were not good fishermen but who owned boats and licences. They had good bankers, but they always presented them with big down payments on deals that were low in risk and low in debt. When he said that, it was clear to me that a lot of small fishermen I know will never get rich because they are afraid of banks. And they are always chasing the big deal that is a gamble.

Many thoughts went racing through my mind as I listened to the Wealthy Fisherman. Rather than making me want to go out and strike it rich, he convinced me that I will never become wealthy. Few fishermen will. But not all wealthy people are happy. So many of them have such a struggle getting rich that they end up as not very nice people. I guess it's because all us poor types are trying to get it away from them.

We all think that if only we could get the secret, we too could become rich beyond our dreams of greed. But like most things I know, in the end there are no secrets. Just common sense and hard work, and sometimes a bit of luck. Let's not kid ourselves—all fishermen believe in luck.

The phone rang. Ed gave me a look. I made my excuses and left. On the drive home I collected my thoughts. Hey, it was fun to meet the Wealthy Fisherman. Then it struck me—what about fun? The books didn't talk about that. I had just learned some secrets about wealth, but surely there was more to it than that. Fun, for example. But what was it? I didn't know. Also, what were Ed and the Wealthy Fisherman up to? That's why I will never be wealthy—I don't know enough.

A Year of Resolution

As Uncle Ed says, predictions are for gypsies. I agree with him. I don't like to predict the future but I also don't like to just sit back and let things happen. And that's why so many of us make New Year's resolutions. I do it every year and for the life of me I can never remember what they were the previous year. The events of the day kind of wipe out what the guiding principles were supposed to be. In spite of all that I am going to make resolutions again this year and this time I'm going to keep them! So this is the year of resolution.

Now the word "resolution" has a special meaning for BC mariners. Captain Cook's ship, the *Resolution*, received her name on December 20, 1771. Cook must have been thinking of the new year also. On one of his great voyages he came to BC on her and even named Ship Cove on Vancouver Island after her. The *Resolution*—a great name for a great ship. So since my inspiration is from a voyage of discovery I resolve to do the same—to discover something new or at least to do something new.

Notes from the Netshed

I think salmon fishing will not be so hot next year, or the year after (I'm not predicting, just stating the facts), so I'll have to make my money from the herring. This year I'm going to resolve to put some money into upgrading my gear. Good gear and preventive maintenance are the best investments you can make. The price of a new outboard is a shock but I resolve to get one for my punt. Next, I will resolve to put together a long-term plan for the herring. Ed and I have wheeled and dealed in that market. We've tried leasing licences, pooling arrangements and anything that looks like it might give us an edge. This year, I will resolve to get a plan. It doesn't include a super-punt—they are fancy, fast and fashionable, but it's the same net that goes in the water, whether it's a super-punt or an oldie like mine. If I do it right I can make some money. My licence would net me a nice nest egg if I cashed in, but right now I would rather it bought me a job, so I resolve to keep it.

I will also resolve to solve the perennial problem of a job on the salmon. I never seem to have one until the last minute. And it's tough to make any decent money. The trend is for less and less fishing time for the fleet. I would like a job on a big seiner as drum man, but without a family connection it's pretty tough to get one of those. I'm tired of crewing on gillnetters and trollers. Being the only woman sharing a small space with a bunch of guys just doesn't compute. I keep resolving to find ways to solve this problem but since I don't have my own boat any more it's out of my hands.

As for the industry, I wish they would resolve to change things. I'm not sure to what, but there has to be a better way of organizing this crazy business. Everybody claims to have the answers but they can't all be right. Maybe all I can do is to resolve not to get upset about it all. There, that fixes the salmon fishery.

A Year of Resolution

There are other fisheries in BC besides the herring and the salmon, but they don't have much appeal for me. I won't dive for geoducks or sea urchins and I can't fish halibut because I have no real experience there. There is supposed to be big money in black cod, but none of the people I know who fish black cod ever seem to have any. So I will resolve not to waste my time trying to get a job there or anywhere else. Although I must admit, I still find these other fisheries interesting enough so that I give them a sideways glance from time to time to see if there is anything new or noteworthy.

Making resolutions about my so-called fishing career is easy because it deals with things that I know and that are concrete. Resolutions about changing my life profoundly are also easy to make, but they are difficult to keep. I know from bitter experience that if I go against my natural instincts I get all screwed up.

So I tend to keep my personal resolutions kind of, well, personal. In modern times it is difficult to be like Captain Cook was in his age. Call your ship *Resolution* and go off and discover Hawaii. Nowadays you have to get clearance for any name you choose for your boat, and you have to have the right gear aboard to travel anywhere, and you can't fish unless there is an opening. But at least we are not cursed with scurvy (again thanks to Cook) and we have flush toilets on fishboats. So our resolutions are perhaps more complex than those that Cook and his pals made.

This sort of thinking comes upon me every year at this time and it can get out of hand. For a good dose of reality I always go over to Ed's place. He's getting older and crankier but still has good advice. You won't find him making resolutions or trying to predict the future. "We'll see" is perhaps his deepest thought. "I've seen a lot of changes and I've been against every one of them," he told me once.

251

Ed is always fascinating. When I told him about my resolutions he said I was all wrong to even attempt to have any. By making resolutions, I'm attempting to control my life in a changing world. Ed lets the world change and then he acts according to his nature. In his opinion, things are in fine shape in the industry. "Otherwise," he snorts, "all of these idiots wouldn't be making a living in it." More boats and fish than ever, Ed points out. Nobody starving. No need for resolutions, you still just need common sense and elbow grease. And so on.

Later I got to thinking about what Ed had said. I would never tell him he was right. That would spoil his fun. But maybe he *is* right. Maybe we don't need to make resolutions to give us a form of Dutch courage to face the New Year.

When I got home I looked up Cook's voyages. It was "written by James Cook, Commander of the *Resolution*." Hey, that was it, Ed was right. It's okay to have resolutions, as long as you are their commander.

AFTERWORD

Mug-Up

As Mrs. Amor de Cosmos illustrates so well, most west coast fisheries have seen major upheavals recently and the future of our fisheries and our fishermen is no longer assured. Disagreements with the US over catch allocations, internal wrangles over shares of the salmon, new licensing regimes and finally El Niño have all contributed to this uncertainty. Once a young person wishing to enter a fishery could easily find a crewing berth and over a few years save enough money to buy a boat and strike out independently, but the cost of doing so in the 1990s has quashed that possibility for most. Licences are worth more than most homes and some are worth many millions of dollars.

Out of the chaos, Mrs. Amor de Cosmos appears to have emerged with some measure of stability. She has a job working as the drum man on a large salmon seiner, and it looks more or less permanent, or as permanent as possible in the fishing industry. She continues to write and catalogue her trials and tribulations. Her Uncle Ed has had a disastrous season and is thinking about quitting fishing as age, rheumatism and crankiness have taken their toll. The Scientific Fisherman endures, trying to make sense out of it all.

The future of the fishing industry and those who make it their life continues to be fraught with political uncertainty. But the wry humour and indomitable spirit that the sea and its fish demand of those who would venture there, will undoubtably see Mrs. Amor de Cosmos, and others like her, through the complex web and lines of the life.

PETER A. ROBSON
EDITOR

Acknowledgments

Good skippers make good fishermen out of deckhands. Good editors sometimes make good writers out of fishermen. I have been fortunate in my editors and I want to thank them here. My first editor in 1981 was Joseph Gough, publisher of the *Canadian Fishing Report* out of Ottawa. There were over fifty "Dear Joe" letters. Joe now works out of the Governor General's office, so he is an inspiration to all who write on fishing matters.

David Rahn of *The Westcoast Fisherman* guided and steered me through some troubled waters starting in March 1991. He saw to it that I never missed a deadline. David is now Group Publisher for Westcoast Publishing and others have taken over his task. Peter A. Robson was my next editor in the magazine and he is doing it again here. His enthusiasm for my writing helped me turn in columns when I sometimes didn't feel like it. Latterly, Michael Skog and Kevin MacDonell have been my editors. Gentlemen, thank you.